# Astronomy For Astro Navigation
(Part of the 'Astro Navigation Demystified' series)

Written by Lt. Cdr Jack Case, M.A. B.Ed.(Hons.)
Published by Bookcase Learning Resources
www.astronavigationdemystified.com
First Edition April 2015
ISBN-13: 9781511675598
ISBN-10: 1511675594

All rights reserved
Registered U.S. Copyright
Registered U.K. Copyright
A catalogue record for this book is available in
The British Library.

**Books in the Astro Navigation Demystified Trilogy:**
Astro Navigation Demystified
Applying Mathematics To Astro Navigation
Astronomy For Astro Navigation

The Astro Navigation Demystified website provides a free resource for all those interested in the subject:
www.astronavigationdemystified.com

# Preface

Astronomy is a vast, complex and very interesting subject; however, when I was taught astro navigation, I often wished that my tutor would 'cut to the chase' and focus on only those aspects of astronomy that were relevant. When I taught navigation myself, I searched high and low for an astronomy book that would allow me to pick out just those topics that I needed to teach my students but I was never able to find one. Throughout my career, I meant to write such a book myself to help other astro navigation tutors and students but I just never found the time. Now, at last, I have had the time to produce the book and I offer it, not just for navigation students but also for practicing navigators. I hope also that teachers in schools and colleges might find it a helpful tool for introducing students to astronomy as well as certain aspects of geography and mathematics.

# About The Author

Jack Case is an experienced navigator who became a teacher when he left the sea. He taught astro navigation, not only to mariners but also to students of mathematics and geography and was able to demonstrate the important links between those subjects. Over many years, he developed the art of teaching navigation in an interesting way so that it could be understood by students of all ages.

For Gwenda, Jade-Marie and Russell.

# Contents

| | |
|---|---|
| **Title Page** | *1* |
| **Preface** | *2* |
| **About The Author** | *3* |
| **Chapter 1** Introduction | *5* |
| **Chapter 2** The Sun and the Earth | *10* |
| **Chapter 3** The Moon | *24* |
| **Chapter 4** The Planets | *34* |
| **Chapter 5** The Stars and Constellations | *47* |
| **Chapter 6** Latitude and Longitude | *114* |
| **Chapter 7** Time | *124* |
| **Chapter 8** Using Celestial Bodies for Navigation | *138* |
| **Chapter 9** Spherical Trigonometry Introduction | *165* |
| **List of Navigational stars** | *169* |
| **Index** | *172* |

# Chapter 1
# Introduction

In coastal waters, we can find our position in relation to geographical features that are marked with pin-point precision on our navigation charts. We do this by triangulation using bearings of the geographical features measured from our position and sometimes combining these bearings with distances, and heights.

When out of sight of land, humans have for centuries, practiced the art of astro navigation (also known as celestial navigation) which involves similar triangulation techniques by measuring the altitudes and azimuths of celestial bodies such as stars, planets, the Sun and the Moon. (The terms azimuth and altitude are explained in chapter 2). The difficulty here of course, is that whereas in terrestrial navigation, the positions of geographical features are fixed and are precisely marked on our charts, the positions of the celestial bodies which are constantly moving through space in different directions and at different speeds, cannot be marked on our charts. So how is this difficulty overcome?

Ptolemy was a Greek astronomer who lived in Alexandria in the first century AD. He believed that the Earth was at the centre of the universe as shown in the following representation of his geocentric model.

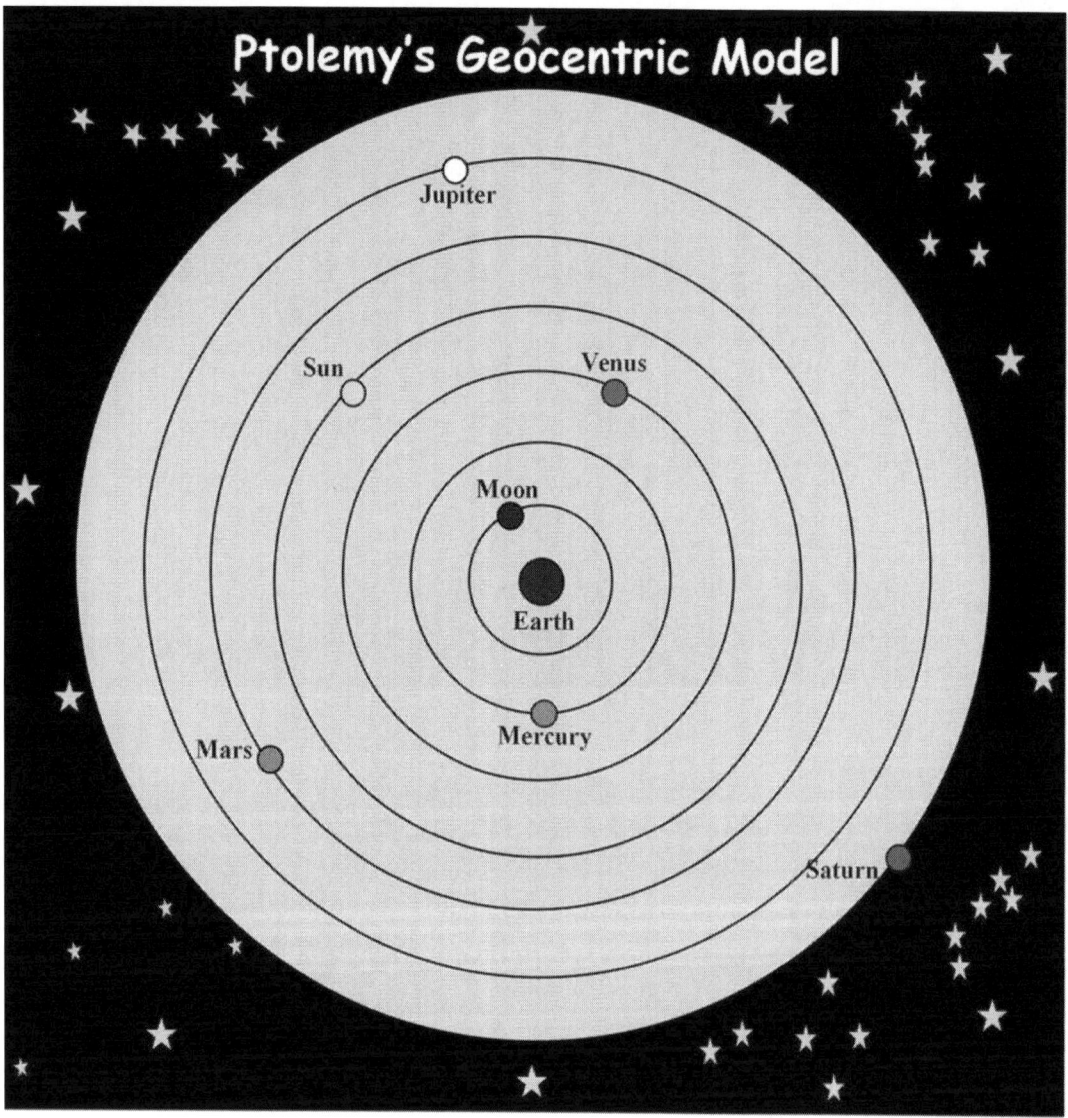

For the purposes of astro navigation, we adopt a similar Ptolemaic worldview. We assume that the Earth is at the centre of a vast sphere which we call the 'Celestial Sphere'. The 'celestial bodies' such as the Sun, Moon, stars and planets are placed on the inner surface of the celestial sphere much as we would see them in the roof of a planetarium.

The fact that the 'celestial bodies' are at greatly varying distances from the Earth and not actually on the inner surface of a sphere is not important since we are only concerned with the angular distances between them. We are not interested in the positions of celestial bodies in terms of astronomical units, we simply need to know how to locate them as they appear to us in the sky.

In the next diagram, X and Y represent two celestial bodies in space and $X^1$ and $Y^1$ represent their corresponding positions on the imaginary celestial sphere.

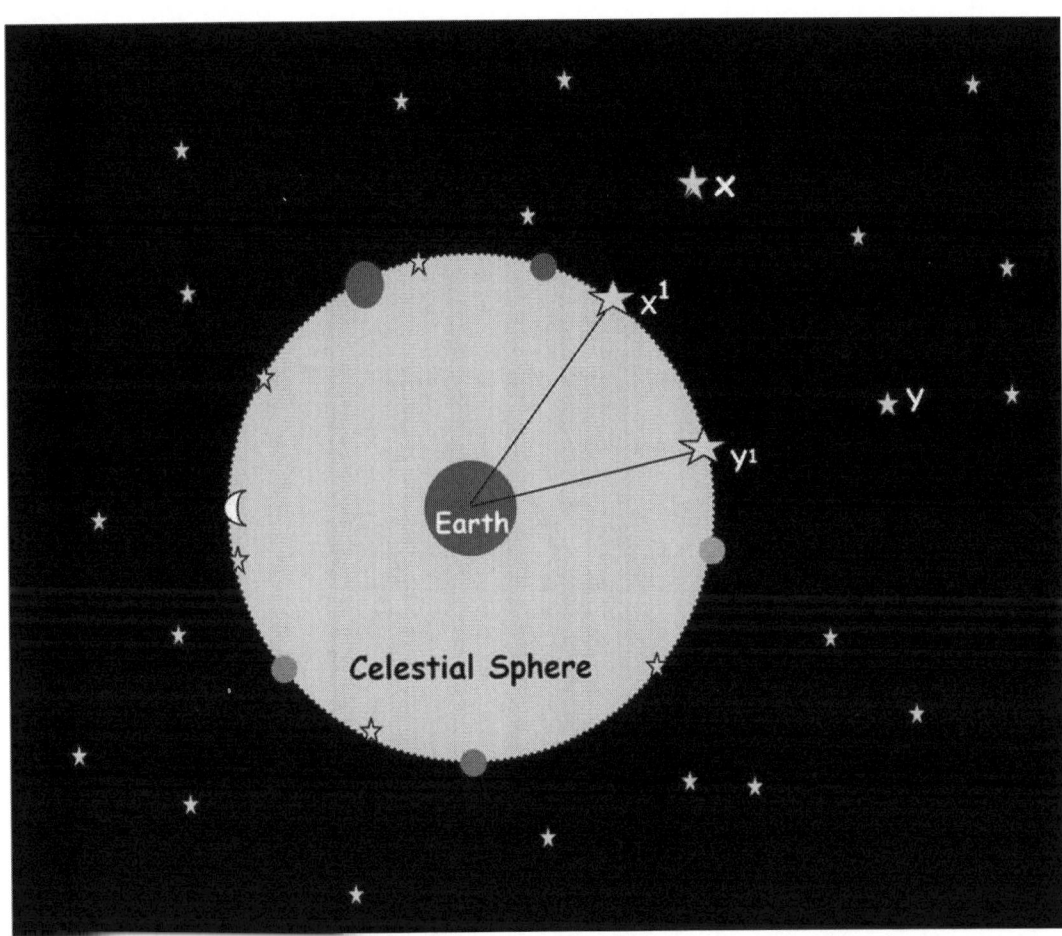

It can be seen that the angular distance between X and Y, when measured from a point on the Earth's surface, is exactly the same as that between $X^1$ and $Y^1$. So, even though the celestial bodies are really at X and Y, no error is introduced by assuming that they are at $X^1$ and $Y^1$.

**Locating Celestial Bodies.** A navigator will use a nautical almanac to find the Greenwich hour angle and the declination of a celestial body and use these to compute its altitude and azimuth at an assumed position through sight reduction methods. There are other methods of determining the positions of celestial bodies such as by the use of Star Globes, Star Charts etc. However, even with these, navigators must eventually locate a body by eye before its altitude and azimuth can be accurately measured at the true position.

**Calculating our position on the Earth's surface by astro navigation.** In the PZX Triangle diagram below, point A represents the position of an observer on the Earth's surface and point Z represents the projection of point A onto the celestial sphere. So Z is the point on the celestial sphere directly above the observer and is called the '**Zenith**

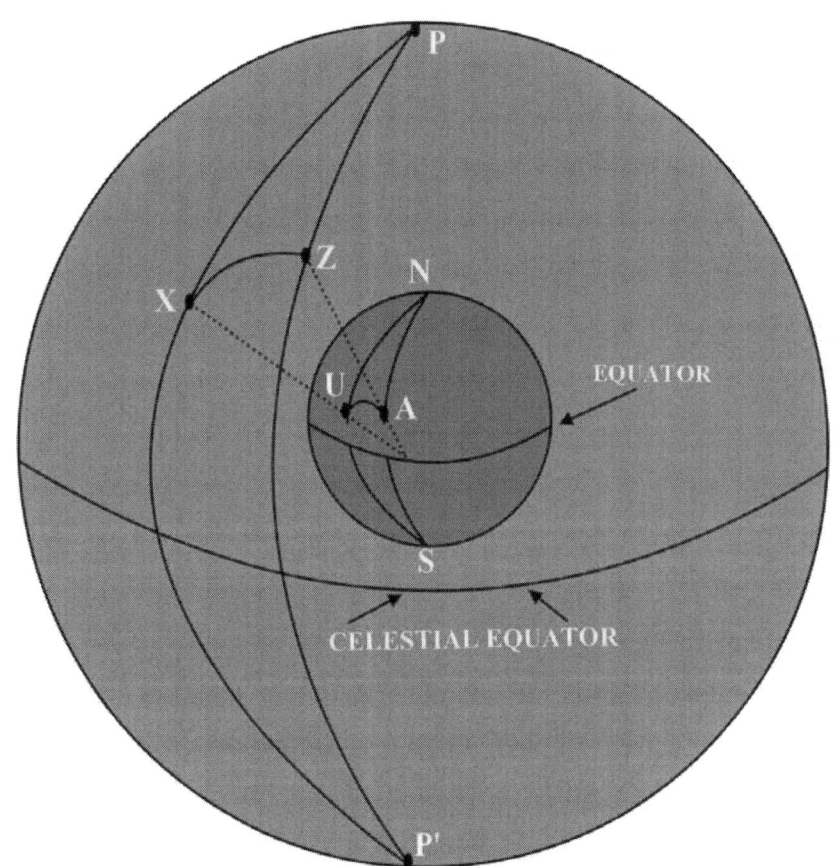

## The PZX Triangle

Point X represents the position of a celestial body and point U represents its projection onto the surface of the Earth. So U is the point on the Earth's surface immediately below the celestial body and is called the 'Geographical Position' (GP).

Similarly, P is the projection of N and P' is the projection of S.

Suppose yourself to be on the Earth's surface at point A. You would not be able to see the North Pole (point N) nor would you be able to see point U (the geographical position). However, you would be able to see the celestial bodies in the sky. So, although the triangle NAU is inaccessible, you would be able to solve it, in effect, by solving the triangle PZX.

As explained in chapter 2, the theory of position fixing by astro navigation depends on the ability to solve the triangle PZX by relating the observed altitude and azimuth of a celestial body as measured at the true position (which is theoretically accurate) to an assumed position (which is only approximate). In this way, we are able to determine the geographical position of the celestial body and then calculate our true position in relation to it. (Note. The book 'Astro Navigation Demystified' which is part of this trilogy, gives a more detailed explanation of the methods used in Astro navigation).

The aim of this book is to look at ways in which this theory can be put to use and we begin by examining the relationships between the Earth and the other celestial bodies.

# Chapter 2
# The Sun and The Earth.

This diagram illustrates the concepts discussed below.

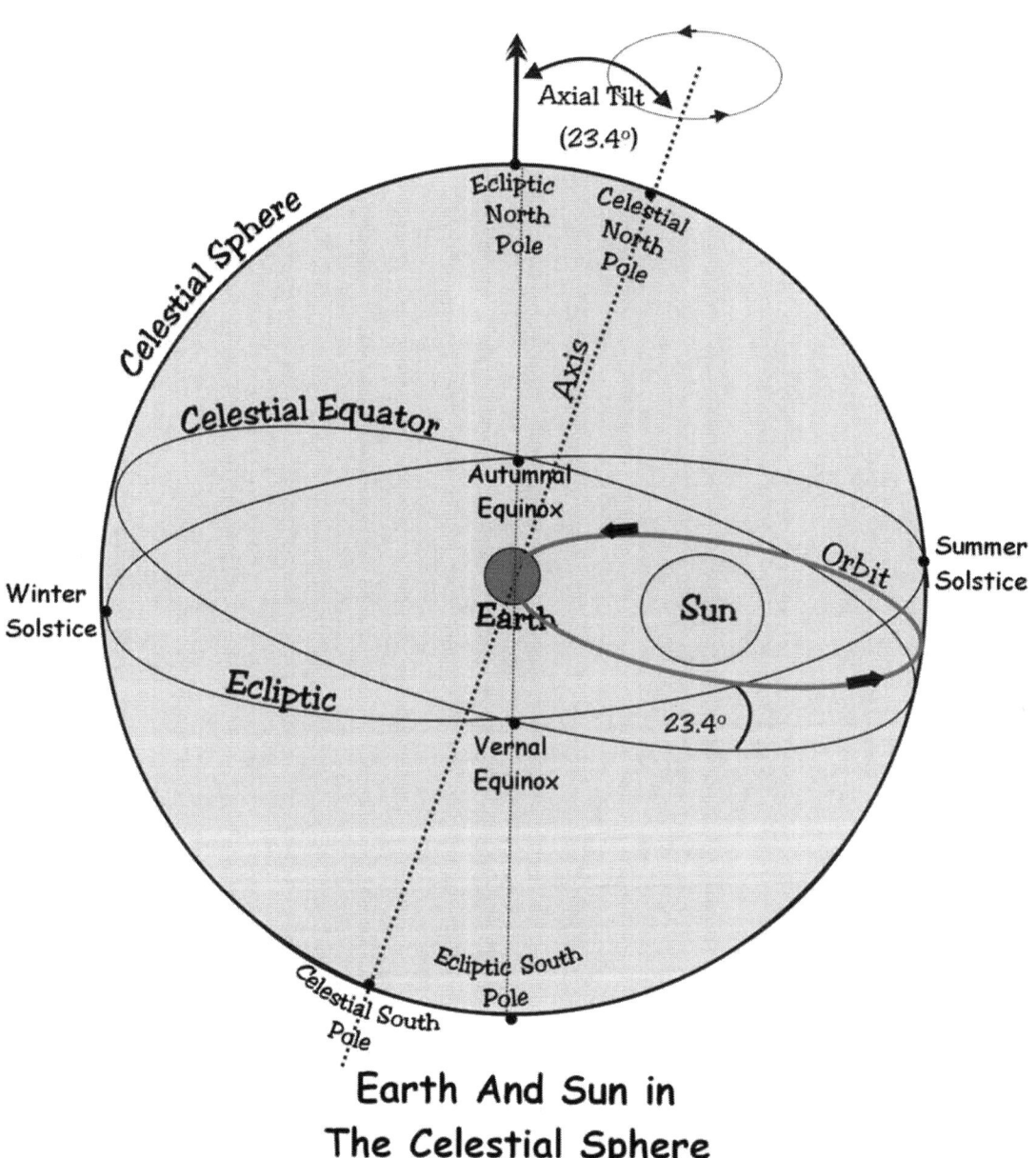

Earth And Sun in The Celestial Sphere

**The Celestial Sphere** is an imaginary sphere with the Earth located at its centre. As discussed in chapter 1, we imagine that the 'celestial bodies' such as the Sun, Moon, stars and planets are placed on the inner surface of the celestial sphere just as we would see them in the sky.

**The Earth.** The Earth is the third planet from the Sun and is the largest of the terrestrial planets which are planets with solid surfaces). (The planets are discussed in greater detail in chapter 4).

### Earth's Orbit.

The Earth's orbit is the elliptical path in which it travels around the Sun. The Earth lies at a mean distance of 93 million miles (149.59787 million kilometers) from the Sun; however, the actual distance varies because of elliptical nature of its orbit. It completes each orbit in 365.256 days (1 sidereal year). The length of the orbit is 940 million kilometers (584 million miles) and although there are slight fluctuations in the orbital speed the average is reckoned to be 30 km/s (67,000 mph).

**Ecliptic.** Because of the orbital motion of the Earth, the Sun appears to us to move around the celestial sphere taking one year to complete a revolution. This apparent movement of the Sun is called the Ecliptic. A year is approximately 365.25 days in length. However; for the sake of convenience, the Gregorian calendar divides three years of the cycle into 365 days and the fourth (the leap year) into 366.

**Ecliptic Poles.** The north and south ecliptic poles are two imaginary points where a straight line drawn from the centre of the Earth and perpendicular to the path of the ecliptic meets the celestial sphere.
**Celestial Poles.** The north and south celestial poles are two imaginary points where the Earth's axis of rotation meets the celestial sphere.
**Geographic Poles.** These are the points where the Earth's axis of rotation meets its surface. These are simply known as the North Pole and the South Pole.
**Magnetic Poles.** These are the north and south poles of the Earth's magnetic field and are offset slightly from the geographical poles.

**Earth's Rotation.** Looking down from the celestial north pole, the Earth rotates about its axis in an anti-clockwise direction or in other words, from west to east. It takes exactly 24 hours for the Earth to turn once on its axis with respect to the Sun but it takes 23 hours, 56 minutes and 4 seconds to complete one rotation with respect to the rest of the universe. The reason for the difference is that, as well as rotating on its axis, the Earth is also orbiting around the Sun and because of this, the Sun catches up with the Earth by approximately 4 minutes each day bringing the rotational period to exactly 24 hours. The amount of time it takes for the Earth to turn on its axis with respect to the universe is know as the sidereal day and the time taken with respect to the Sun is called a solar day.

**Axial Tilt.** This is the angle between the Earth's rotational axis and a line perpendicular to the ecliptic which passes through the geocentric centre of the Earth and its celestial poles. Axial tilt is measured between the celestial and the ecliptic poles. The Earth currently has an axial tilt of 23.4°; however, this is not a fixed quantity and at present, it is decreasing at a rate of about 47 arc seconds per century. This is due to a gravity induced gradual shift in the orientation of the Earth's rotational axis and is known as **Axial Precession**.

**Rotational Velocity.** The Earth's rotational velocity at the equator is 1,674.4 km/h. so a person standing on the equator would be travelling 1,674.4 km/h in a circle.

**The Earth's equator** is an imaginary line on the Earth's surface the plane of which is at right angles to the axis of rotation. It is equidistant from the North and South Poles and divides the Earth into the Northern Hemisphere and Southern Hemisphere. The length of the equator (that is the equatorial circumference) is 40,075.16 km. which is equal to 21600.5 geographical miles.

**The Celestial Equator** is the projection of the Earth's equator onto the surface of the celestial sphere.

**Declination.** The declination of a celestial body is its angular distance North or South of the Celestial Equator. The declinations of the stars change very slowly and can be considered to be almost constant for up to a month at a time.
The declination of the Sun changes relatively fast from 23.4° North to 23.4° South and back again during the course of a year.
The Moon's declination is more difficult to predict because the rate of change is even more rapid than that of the Sun and the pattern of the changes is less uniform.
Like the Sun and the Moon, the declinations of the planets also change rapidly in comparison with the stars.
Declination can be summarised as the celestial equivalent of Latitude and for practical reasons, we treat it as the angular distance of a celestial body North or South of the Equator.

**The Equinoxes.** The Sun crosses the celestial equator on two occasions during the course of a year and these occasions are known as the equinoxes. At the equinoxes, at all places on Earth, the nights and days are of equal duration (i.e. 12 hours) hence the term equinoxes (equal nights). Because the Sun is on the celestial equator at the equinoxes, its declination is of course 0°.
**The Autumnal Equinox** occurs on about the 22nd. September when the Sun crosses the celestial equator as it moves southwards from 23.4°N, the northernmost limit of its declination.

**The Vernal Equinox** occurs on about the 20th March when the Sun crosses the celestial equator as it moves northwards from 23.4°S, the southernmost limit of its declination.

**The Solstices.** The times when the Sun reaches the northerly and southerly limits of its path along the ecliptic are known as the solstices. The word solstice is taken from 'solstitium', the Latin for 'sun stands still'. This is because the apparent movement of the Sun seems to stop before it changes direction

**The Summer Solstice** (mid-summer in the northern hemisphere) occurs on about 21st June when the Sun's declination reaches 23.4° North (the tropic of Cancer).

**The Winter Solstice** (mid-winter in the northern hemisphere) occurs on about 21st December when the Sun's declination is 23.4°South (the tropic of Capricorn).

**Note.** Because each year is 365.25 days in length, the dates of the equinoxes and the solstices will vary slightly during the four-year cycle between leap years; so the Vernal Equinox sometimes falls on 20th March and sometimes on 21st. The Autumnal Equinox sometimes falls on 22nd September and sometimes on 23rd. Similarly, the Summer Solstice usually falls on 21st June but sometimes falls on 20th. The Winter Solstice usually falls on 21st December but sometimes falls on 22nd.

**The tropic of Capricorn** is so named because in ancient times, the Sun passed through the constellation Capricornus during the Winter Solstice on 21/22 December when the Sun's declination reached its southernmost latitude of 23.4°S. However, due to precession, the Sun is now over the constellation Sagittarius at the Winter Solstice.

**The Tropic of Cancer.** These days, the Sun passes through the constellation Cancer in late July; however, in the time of Ptolemy, around 2000 years ago, this occurred during the summer solstice when the Sun reached 23.4° N, the northern limit of the ecliptic. The latitude 23.4°N is still called the tropic of Cancer even though the Sun now resides in Taurus at the summer solstice.

**Note.** The latitude of the tropic of Cancer is currently drifting south at approximately 0.5 arc seconds per year while the latitude of the tropic of Capricorn is drifting north at the same rate.

**Zenith.** The **Zenith** is an imaginary point on the celestial sphere directly above the observer. It is the point where a straight line drawn from the geocentric centre of the Earth, through the observer's position and onwards, intersects with the celestial sphere.

**Zenith Distance** is the angular distance from the zenith to the celestial body measured from the Earth's centre.

**Nadir.** The Nadir is the direction opposite to the zenith, that is the direction pointing directly below the observer.

**Altitude and Azimuth**

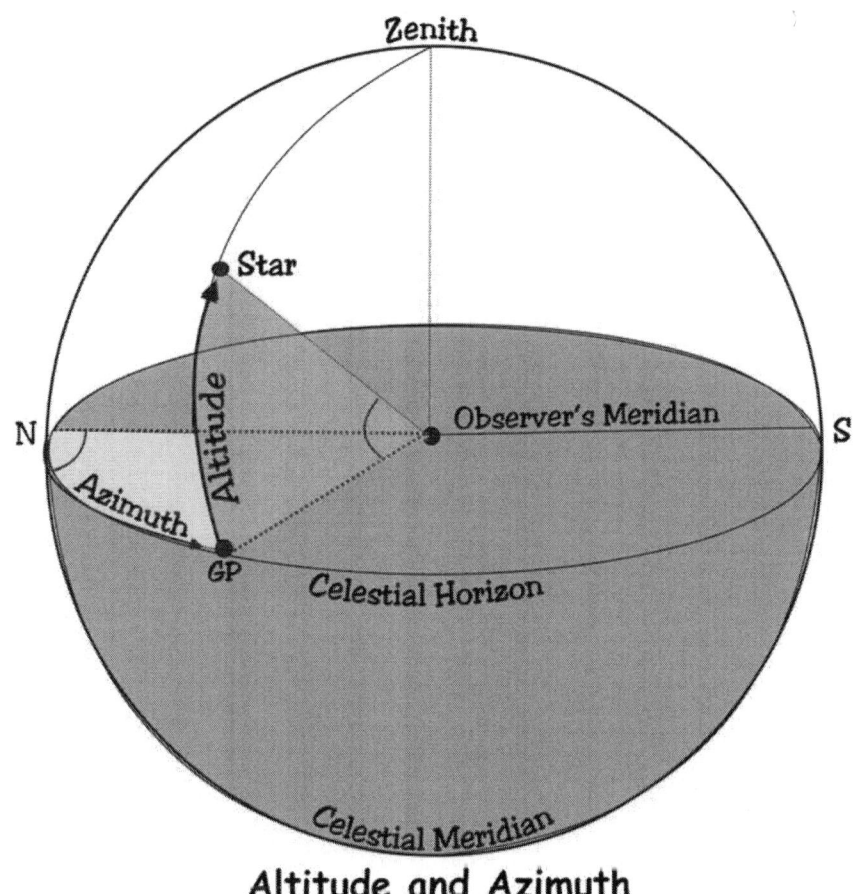

**Altitude and Azimuth**

**Altitude.** The altitude of a celestial body is the angular distance between its position in the celestial sphere and the celestial horizon as measured at the observer's position.

**Azimuth.** The azimuth of a celestial body is the angular distance between the observer's meridian and the direction of the geographical position of the body (GP) measured at the pole (north pole in the northern hemisphere and the south pole in the southern hemisphere).

**Visible Horizon (also called the Sensible Horizon).** The plane of a small circle of the celestial horizon which is perpendicular to the zenith of the observer's position is called the visible horizon. In other words, it is the limit of the horizontal plane at the observer's position.

**Celestial Horizon (also called the Rational Horizon).** The celestial horizon is the plane of a great circle that passes through the Earth's centre and is parallel to the visible horizon.

**Observed Altitude and True Altitude.** As shown in the next diagram, the observer measures the altitude in relation to the visible horizon from his position at O on the Earth's surface. So, the **observed altitude** is the angle HOX. However, the **true altitude** is measured from the Earth's centre in relation to the celestial horizon and is the angle RCX.

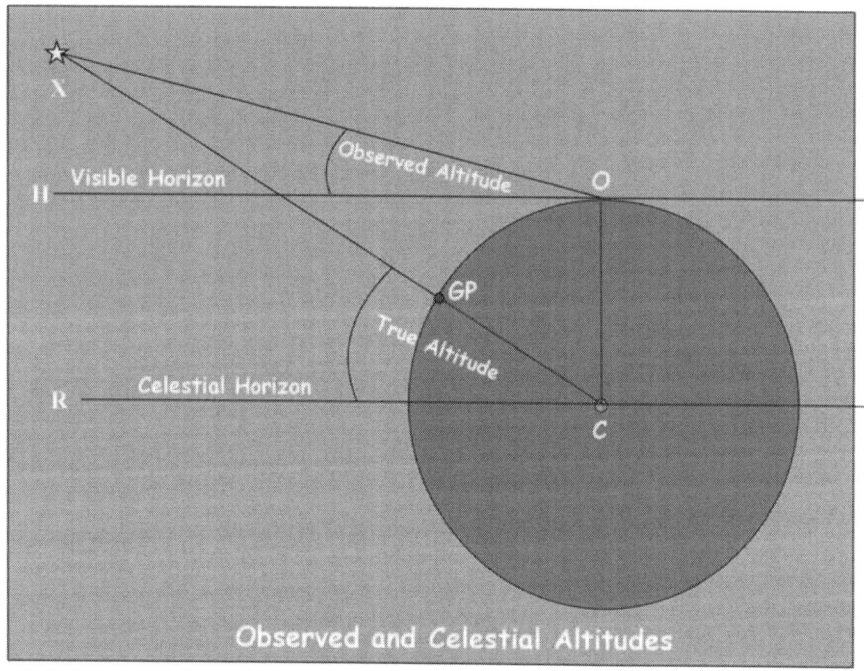

Observed and Celestial Altitudes

**Local Hour Angle (LHA).** In astro navigation, we need to know the position of a celestial body relative to our own position.

*(The following explanation refers to the PZX triangle diagram which is repeated below).*

LHA is the angle ANU on the Earth's surface which corresponds to the angle ZPX in the Celestial sphere. In other words, it is the angle between the meridian of the observer and the meridian of the geographical position of the celestial body (GP).

Due to the Earth's rotation, the Sun moves through 15° of longitude in 1 hour and it moves through 15 minutes of arc in 1 minute of mean time. So the angle ZPX can be measured in terms of time and for this reason, it is known as the Local Hour Angle.

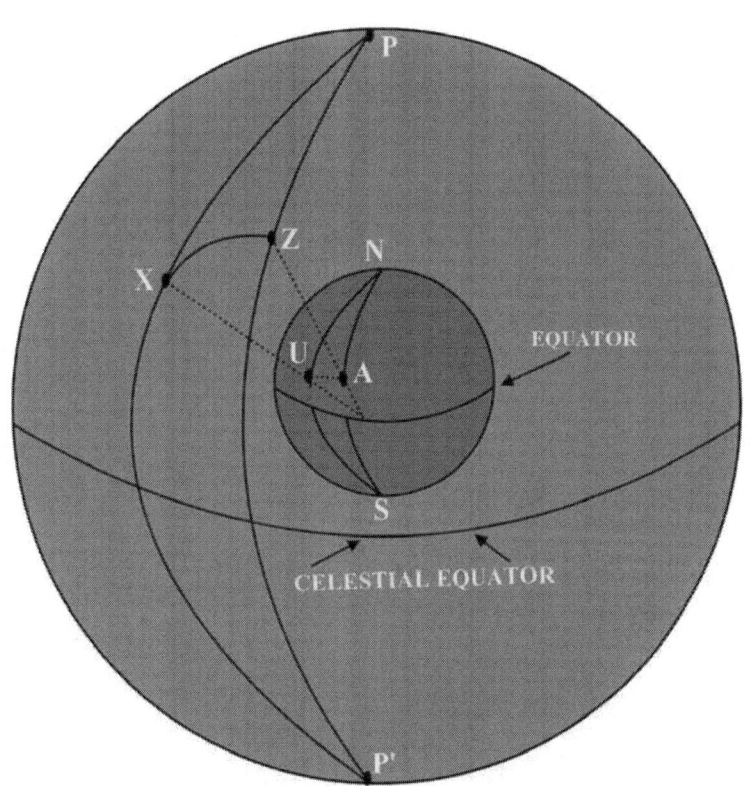

**The PZX Triangle**

LHA is measured westwards from the observer's meridian and can be expressed in terms of either angular distance or time. For example, at noon (GMT) the Sun's GP will be on the Greenwich Meridian (0°). If the time at an observer's position is 2 hours and 3 minutes after noon, then the angular distance between the observer's meridian of longitude and the Greenwich Meridian must be (2 x15°) + (3x 15') = 30° 45'.

Because it is after noon at the observer's position, the longitude of that position must be to the East of the Greenwich Meridian since the Earth rotates from west to east. Therefore, the observer's longitude must be 30° 45' East and since LHA is measured westwards from the observer's meridian, the LHA must also be 30° 45'. However, it should be noted that as the Earth continues to rotate eastwards, the GP of the Sun will continue to move westwards so the LHA at the observer's position will be continually changing.

**Greenwich Hour Angle (GHA).** As discussed above, the angle between two meridians of Longitude can be expressed as an hour angle. The hour angle between the Greenwich Meridian and the meridian of a celestial body is known as the Greenwich Hour Angle.

The Local Hour Angle between an observer's position and the geographical position of a celestial body can be found by combining the observer's longitude with the GHA. This is demonstrated in the following diagram.

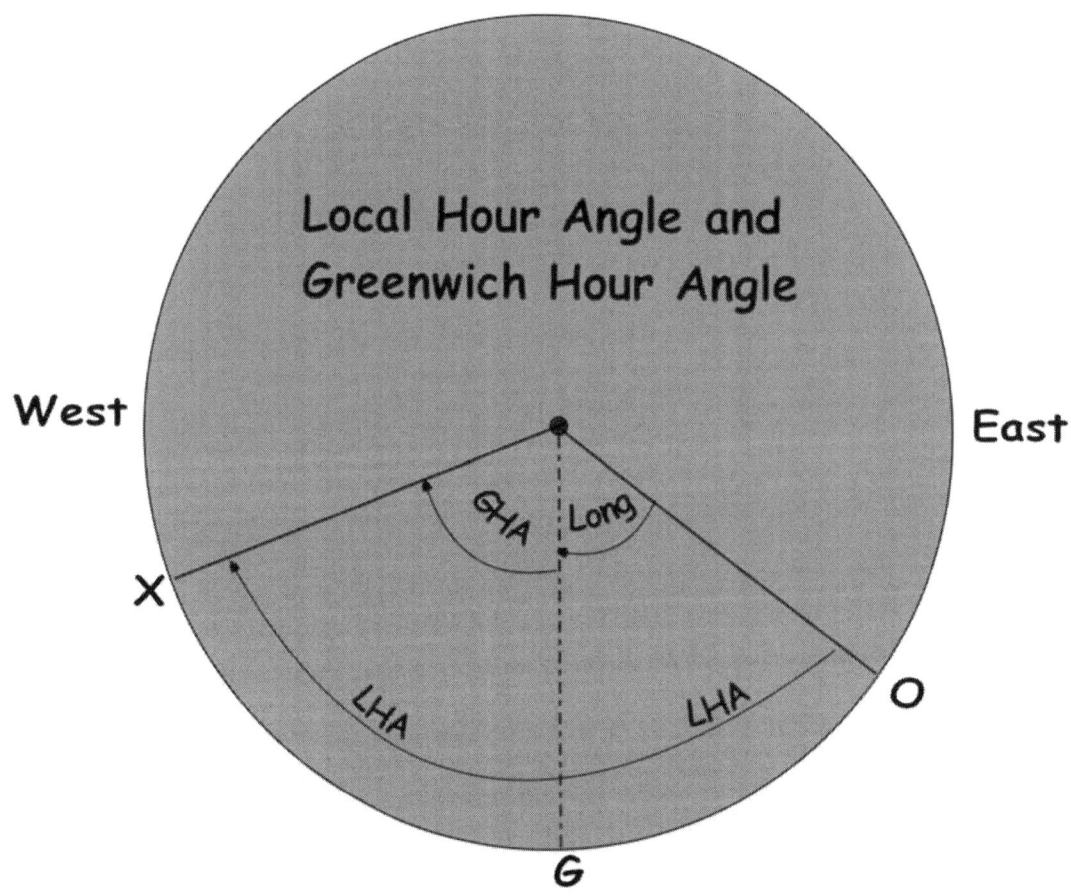

In the diagram, O represents the longitude of an observer;
X represents the meridian of a celestial body;
G represents the Greenwich Meridian.
Because, in this case, the observer's longitude is east and because LHA is measured westwards from the observer's meridian to the meridian of the celestial body, LHA is equal to the GHA plus the longitude. If the longitude were to be west then this rule would change so that LHA would equal GHA minus Long.
(Note. This topic is covered in greater depth in the book 'Astro Navigation Demystified').

**First Point of Aries.** In astronomy, we need a celestial coordinate system for fixing the positions of all celestial bodies in the celestial sphere. To this end, we express a celestial body's position in the celestial sphere in relation to its angular distances from the Celestial Equator and the celestial meridian that passes through the **'First Point of Aries'**. This is similar to the way in which we use latitude and longitude to identify a position on the Earth's surface in relation to its angular distances from the Equator and the Greenwich Meridian.
The First Point of Aries is usually represented by the 'ram's horn' symbol shown below:

Just as the Greenwich meridian has been arbitrarily chosen as the zero point for measuring longitude on the surface of the Earth, the first point of Aries has been chosen as the zero point in the celestial sphere. It is the point at which the Sun crosses the celestial equator moving from south to north along the ecliptic (at the vernal Equinox in other words). This point is known as the 'First Point of Aries' because in 150 B.C. when Ptolemy first mapped the constellations, Aries lay in that position. However, although still named the 'first point of Aries', due to precession, the vernal equinox now lays in the constellation Pisces.

**Right Ascension (RA).** This is used by astronomers to define the position of a celestial body and is defined as the angle between the meridian of the First Point of Aries and the meridian of the celestial body measured in an

Easterly direction from Aries. RA is not used in astro navigation; Sidereal Hour Angle is used instead:

**Sidereal Hour Angle (SHA).** This is similar to RA in as much that it is defined as the angle between the meridian of the First Point of Aries and the meridian of the celestial body. However, the difference is that SHA is measured westwards from Aries while RA is measured eastwards.

The following diagram illustrates the concepts discussed above.

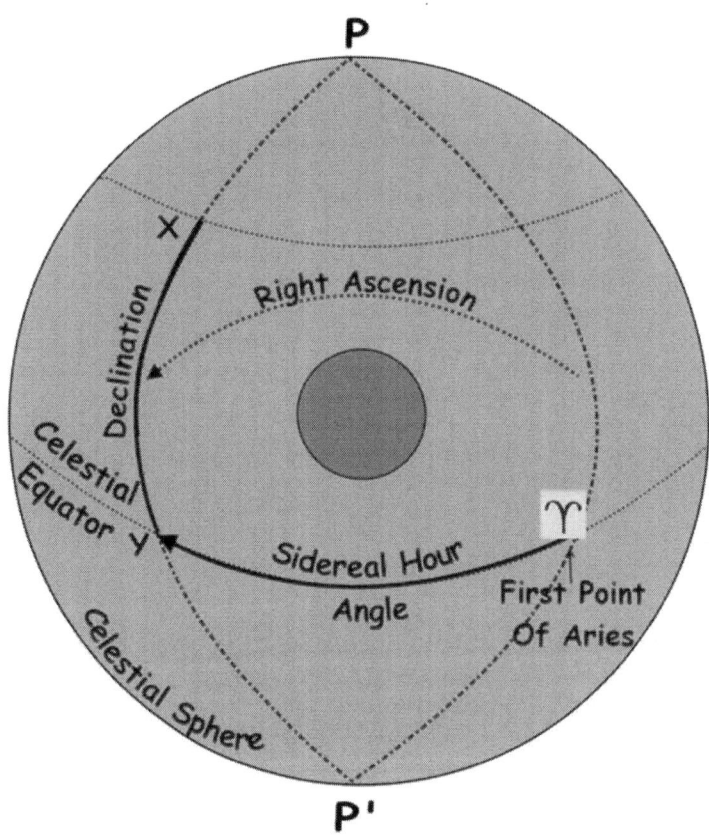

Sidereal Hour Angle, Right Ascension
and The First Point Of Aries

X is the position of a celestial body in the celestial sphere.
PXP' is the meridian of the celestial body.
Y is the point at which the body's meridian crosses the celestial equator.
♈ is the First Point of Aries.

The **Sidereal Hour Angle** is the angle ΥPY. That is the angle between the meridian running through the First Point of Aries and the meridian running through the celestial body measured at the pole P.

It can also be defined as the angular distance ΥY. That is the angular distance measured **westwards** along the Celestial Equator from the meridian of the First Point of Aries to the meridian of the celestial body.

**Right Ascension** can also be defined as the angle between the meridian of the First Point of Aries and the meridian of the celestial body but the difference is that it is measured in an **easterly** direction from Aries.

From this, we can conclude that

RA  = 360° - SHA and
SHA = 360° - RA.

**Astro Navigation – the Importance of Altitude, Azimuth and Zenith Distance.** Please consider the following diagram. The celestial sphere is drawn in the plane of the observer's meridian with the observer's zenith (Z) at the top.

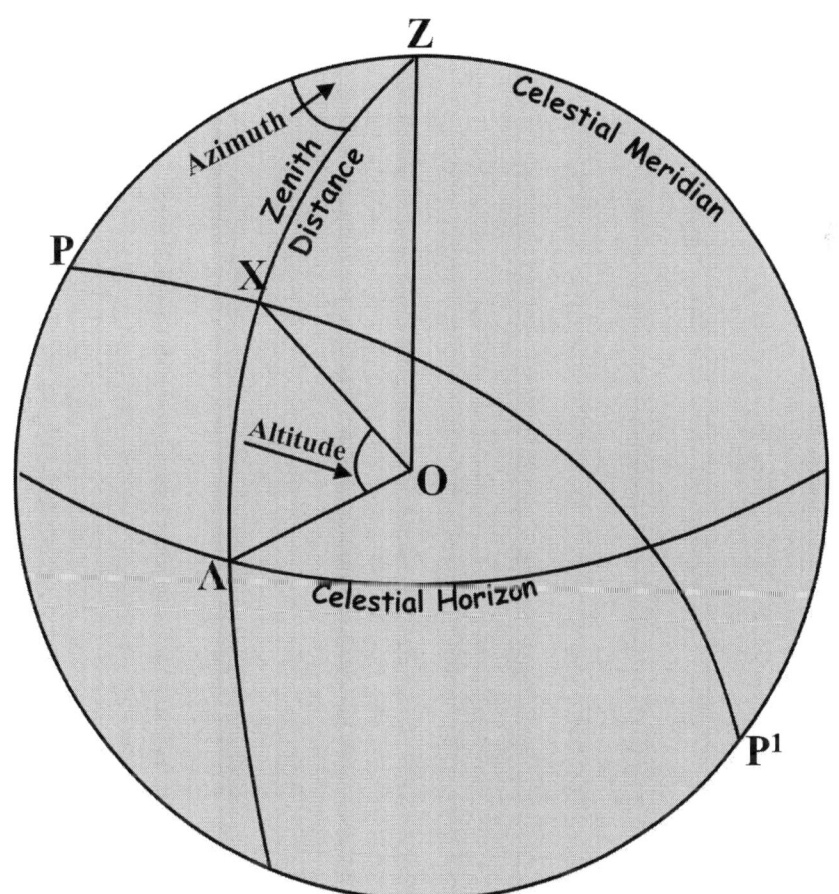

**Altitude, Azimuth and Zenith Distance**

Point O represents both the observer and the Earth.
Z represents the observer's zenith.
X is the position of a celestial body in the celestial sphere.
A is the point where the virtual circle running through the position of the celestial body meets the celestial horizon.
P and $P^1$ are the north and south poles respectively.

**The Zenith Distance.** In this diagram, the zenith distance is the angular distance ZX that is subtended by the angle XOZ and is measured along the vertical circle that passes through the celestial body. (A vertical circle is a great circle that passes through the observer's zenith and is perpendicular to the celestial horizon).

**The Altitude.** Altitude is the angle AOX, that is the angle from the celestial horizon to the celestial body and is measured along the same vertical circle as the zenith distance.

**Relationship Between Zenith Distance And The Nautical Mile.** An angle of 1 minute at the earth's centre will subtend an arc of length 1 n.m on the earth's surface. Therefore if the angle XOZ is 30° (the arc ZX) will be equal to 30 x 60 = 1800 arc minutes at the earth's surface and so the zenith distance will be equal to 1800 nautical miles.

**Relationship between Altitude and Zenith Distance**
Since the celestial meridian is another vertical circle and is therefore, also perpendicular to the celestial horizon, it follows that angle AOZ is a right angle and angles AOX and XOZ are complementary angles. From this we can deduce that:
**Zenith Distance = 90° – Altitude**
and **Altitude = 90° – Zenith Distance**

At this point, we need to take another look at the PZX Triangle diagram which is repeated below.
In this diagram the arc AU is the arc joining the observer's position to the geographical position of the celestial body. This arc when projected onto the celestial sphere forms the arc ZX which is the zenith distance. Therefore, from the discussion above, it can be seen that the angular distance ZX is equal to the angular distance AU which when converted to

nautical miles will give us the distance from the GP of the body to the position of the observer.

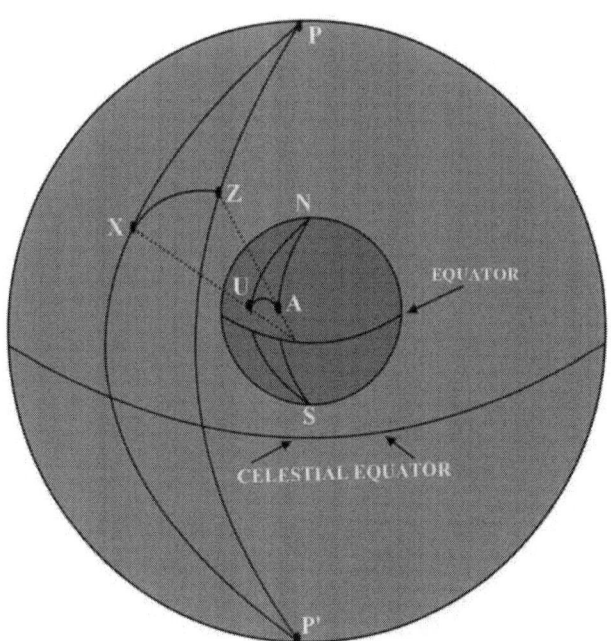

The PZX Triangle

**Azimuth.** The angle PZX is the azimuth of the celestial body and is the angular distance between the observer's celestial meridian and the direction of the position of the body. It is similar to the bearing except that whereas bearing is always measured from 0° to 360° clockwise from the north pole, azimuth is measured from 0° to 180° east or west from the north pole in the northern hemisphere and from the south pole in the southern hemisphere.

**Summarizing The Role Of Altitude, Azimuth And Zenith Distance In Astro Navigation.** The preceding discussion illustrates the importance of altitude and azimuth in astro navigation. It can be seen that by measuring the altitude of a celestial body, we are able to easily calculate the zenith distance which will give us the distance in nautical miles from the observer's position to the geographical position of the body. The azimuth will give us the direction of the GP from the observer's position. This explains why measuring the altitude and azimuth are the first steps in determining our position in astro navigation. (A thorough treatment of this topic can be found in the book Astro Navigation Demystified).

# Chapter 3
# The Moon

The Moon which is our only natural satellite completes its orbit in 27.3 Earth days. It has a circumference of 10,917 km. and is 27% the size of the Earth. Its average distance from the Earth is 384,400 Km. but it is drifting away at approximately 3.7 cm. per year.

**The Dark Side of the Moon.** Because the Moon rotates around its own axis at the same time that it takes to orbit the Earth, the same side always faces the Earth and is lit by sunlight. For that reason, we can only ever see one side of the Moon from the surface of the Earth. The other side of the Moon is in perpetual darkness and was not seen by humans until 1968 when Jim Lovell, Frank Borman and William Anders went into orbit around it in the Apollo 8.

# PHASES OF THE MOON

The diagram below shows, that as the Moon completes its 27.3 day orbit around the Earth, we see it pass through various phases of illumination. It goes from New Moon, to Full Moon and back to new Moon again.

**The Phases.**
**New Moon.** When the illuminated side of the Moon is facing away from the Earth. The Moon and the Sun are lined up on the same side of the Earth, so we can only see the shadowed side.
During a new moon, we can see the reflected light from the Earth, since no sunlight is falling on the Moon – this is known as earthshine.

**Waxing Crescent** – The waxing crescent moon is the first sliver of the Moon that we can see after the new moon. From the northern hemisphere, the crescent moon has the illuminated edge of the Moon on the right. This situation is reversed for the southern hemisphere. "Waxing" means that the Moon becomes more illuminated night-by-night,

**First Quarter** – This occurs when the Sun and the Moon make a 90-degree angle compared to the Earth. Although it's called a quarter moon, we actually see it as half illuminated.

**Waxing Gibbous** – This phase of the Moon occurs when more than half of the Moon is illuminated but it is not yet a full Moon.

**Full Moon** – This is the phase when the Moon is brightest in the sky. The Moon and the Sun are lined up on opposite sides of the Earth, so from our perspective here on Earth, the Moon is fully illuminated by the light of the Sun.

**Waning Gibbous** – In this lunar phase, more than half of the Moon is illuminated but it is not yet a full moon. The term "waning" means that it is getting less illuminated each night.

**Last Quarter** – At this point of the lunar cycle, the Moon has reached half illumination again. Now it is the left-hand side of the Moon that's illuminated, and the right-hand side is in darkness (from a northern hemisphere perspective).

**Waning Crescent** – This is the final sliver of illuminated moon we can see before the Moon goes into darkness again.

## The Ocean Tides.

The rise and fall of the ocean tides is caused by the gravitational forces of the Sun and the Moon.

**Tidal Effects of the Moon.** If it were not for the gravitational attraction of the Sun and the Moon, the water level of the seas and oceans would be kept at equal levels by a combination of the Earth's own gravity pulling it inwards and centrifugal force pushing it outwards. However, the gravitational force of the Moon is strong enough to attract the water towards it and cause a bulge beneath it. As the Earth rotates and the Moon orbits around it, the bulge follows the Moon causing high tides in its vicinity.

The combined effects of the Earth's rotation and the Moon's orbit around it cause the 'bulge' to move around the Earth in 24 hours and 50 minutes and so it would seem, at first sight, that we would get high tides only at that time interval. However, other forces are at play. On the opposite side of the Earth to where the 'bulge' occurs, the Moon's gravitational pull is at its weakest and this allows the Earth's centrifugal force to push the water outwards and so cause another bulge therefore giving us two high tides a day. This means that the time between high tides is approximately 12 hours and 25 minutes and the time between high tide and low tide is 6 hours 12.5 minutes in deep ocean areas. This can change dramatically owing to a variety of factors such as the topography of the ocean floor, local currents, varying water depths and the declination of the Moon.

The height of the tides vary during the course of a month because the Moon is not always at the same distance from the Earth due to its elliptical orbit. As the Moon comes closer to the planet, its gravitational pull increases and this leads to higher tide levels. Conversely, when the Moon's orbit takes it further away from the Earth, the tides become lower. When the Moon is at its closest distance to the Earth, its gravitational pull increases by as much as 50% and this leads to higher sea levels on Earth. When it is at its furthest distance, sea levels are much lower,

**Tidal Effects of the Sun.** The Sun also affects the rise and fall of the tides on Earth. The gravitational attraction of the Sun pulls the ocean water towards it but at the same time, the effect of the Earth's rotation around the Sun creates a centrifugal force which pushes the water outwards on the side facing away from the Sun. The combined effect of these two forces creates a tidal bulge on the side of the Earth facing away from the Sun. However, this effect of the Sun is less than that created by the Moon which is much closer to the Earth.

**Spring Tides.** When the Sun, the Moon and the Earth are in syzygy, that is when they are lined up as during a Full Moon or New Moon, the combined tidal effect of the Sun and Moon is at its greatest and causes what is known as Spring Tides. This has nothing to do with the season of Spring but to do with the saying that the water 'springs' higher than normal.

**Neap Tides.** When the directions of the Sun and the Moon in relation to the Earth are at right angles, as during a Quarter Moon, the combined effects of their gravitational pull is less and so the height of the tides is much lower and are known as Neap Tides.

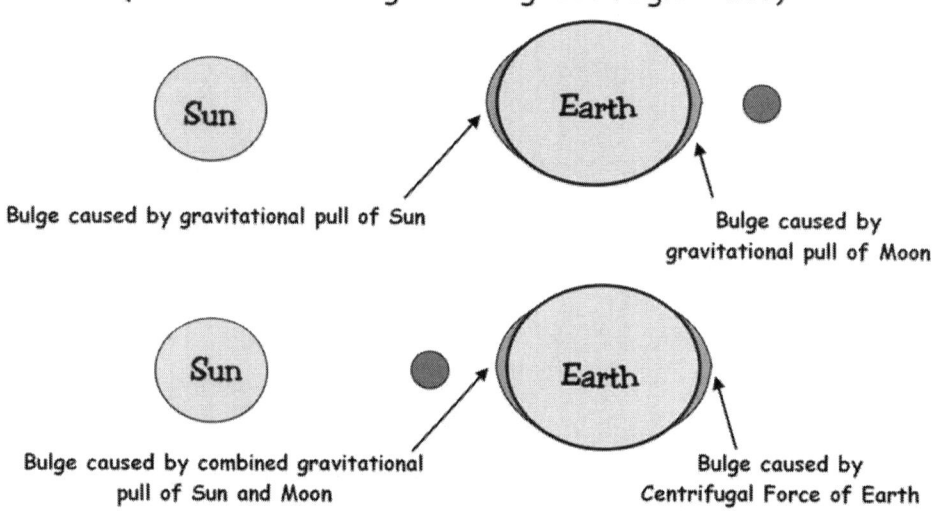

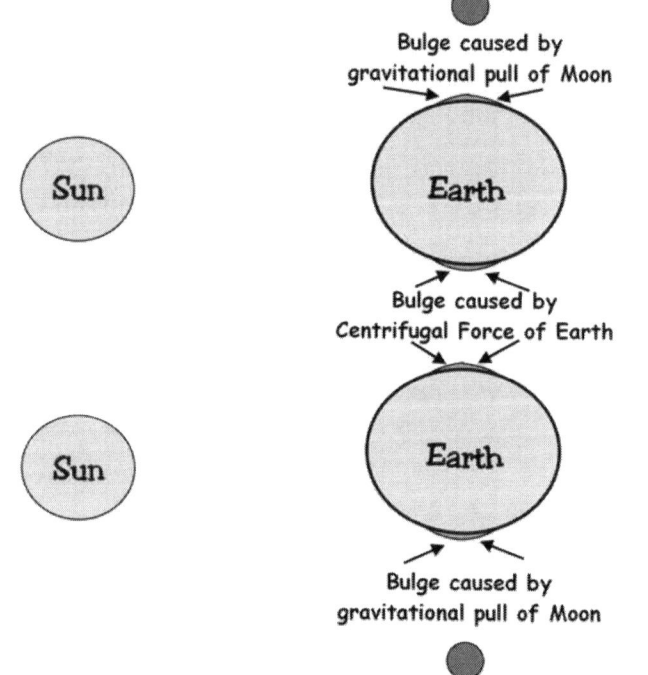

**Measuring the Altitude of the Moon.**   When a navigator measures the altitude of the Moon, there are several corrections that he has to make to the readings.

**Corrections For The Moon's Semi-Diameter.**  The point on the Moon's circumference nearest to the horizon is called the **lower limb** and the point furthest from the horizon is called the **upper limb**.
When the Moon is not full, sometimes only the upper limb will be visible and sometimes only the lower limb.
From the diagram below it can be seen that sometimes, depending on the phase of the Moon, either the upper or the lower limb cannot be seen.

Waxing Crescent Moon    Waxing Gibbous Moon

Waning Crescent Moon    Waning Gibbous Moon

It should be noted that whether the Moon's upper or lower limb is visible is dependent not only on its phase but also on the relative altitudes of the Sun and the Moon.  For example, if, one morning, a crescent or gibbous moon is visible in the eastern sky and the Sun is at a higher altitude, only the upper limb will be visible but if, in the evening of the same day, the Moon is visible in the western sky and the Sun has set below the western horizon, only the lower limb will be visible.

In navigational practice, the altitude that we measure is that of the lower limb; however, when the lower limb cannot be seen, we have no choice other than to measure the altitude of the upper limb.

Regardless of which limb we use, what we really need is the altitude of the Moon's centre so we must either add or subtract the value of its semi-diameter.
The following diagram shows why the semi-diameter must be added when the altitude when the lower limb is measured.

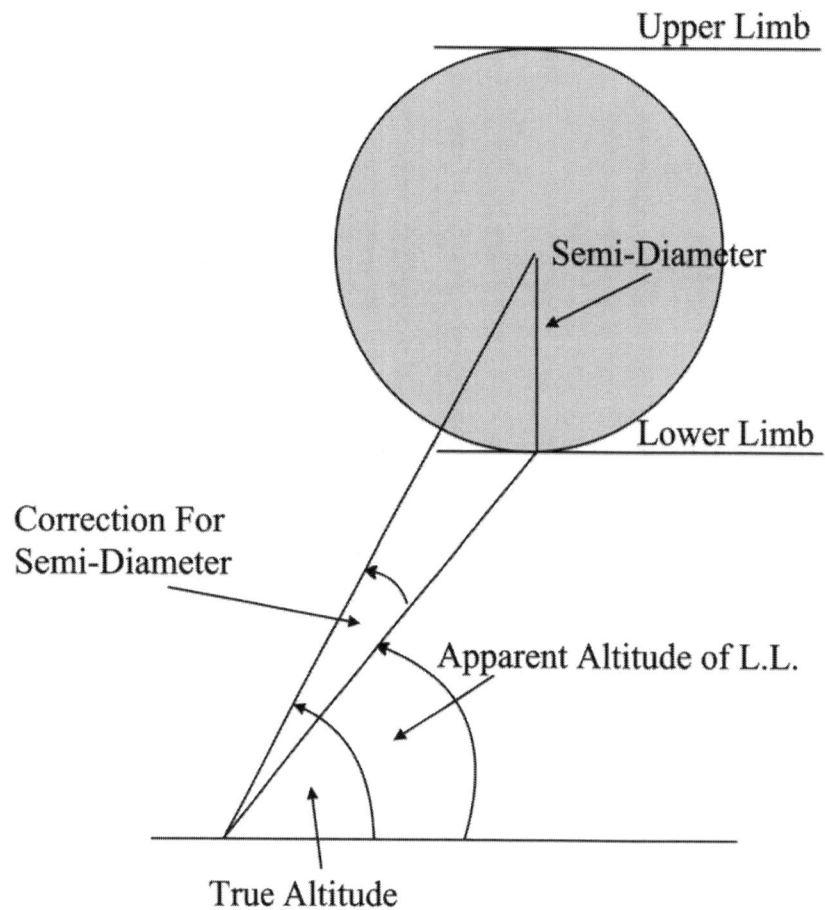

Semi-Diameter From Lower Limb

As the Moon travels around its orbit and its distance from the Earth changes, so the value of the visible moon's semi-diameter will change. The value of the Moon's semi-diameter for each day is given in the daily pages of the Nautical Almanac.

**Corrections For Refraction.** When a ray of light from a celestial body passes through the Earth's atmosphere, it becomes bent through refraction and this causes the apparent (observed) altitude to be greater than the true altitude. Since the sextant measures the apparent altitude, a correction for refraction must be applied to find the true altitude. Refraction is at its greatest when the altitude is small (i.e. when the celestial body is near the horizon) and becomes less as the altitude increases.
The effects of refraction are illustrated in the diagram below.

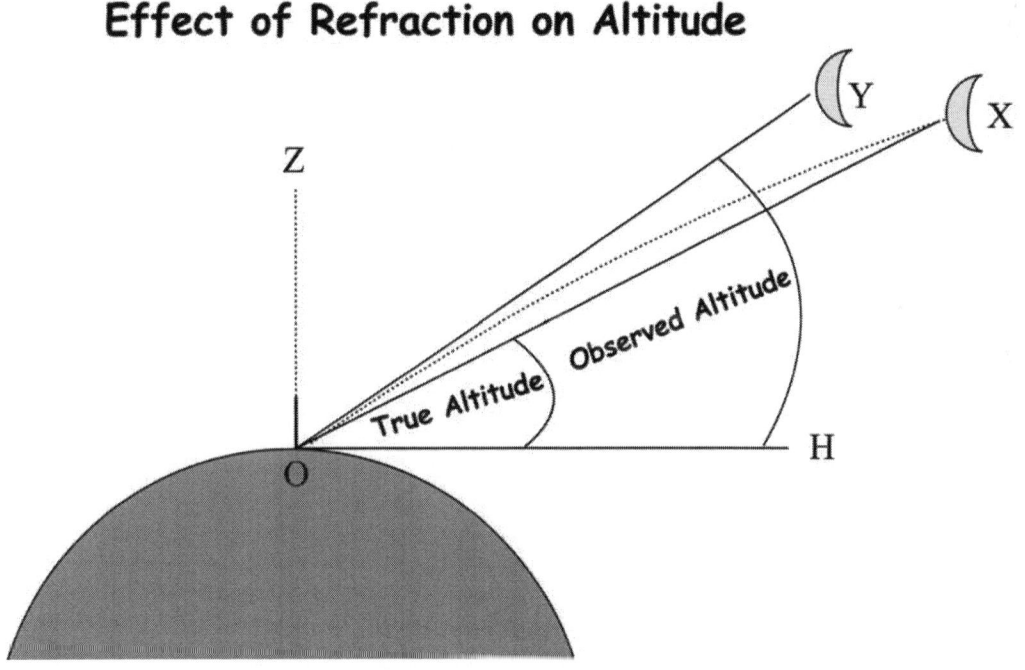

O is the observer's position and Z is the zenith at that point.
OH is the horizon
XOH is the true altitude of the Moon from the observer's position.
However, due to refraction, the celestial body appears to be at Y and so YOH becomes the observed altitude and a correction will have to be made to compensate for this.

**Corrections For Parallax.** We measure the altitude of a celestial body from our position in relation to our visible horizon; this is known as the **observed altitude**. However, when calculating the **true altitude**, measurements are made from the Earth's centre in relation to the celestial horizon. The displacement between the observed position of an object and the true position is known as **parallax**.

**Parallax corrections for the Moon.** Because the Sun and the Moon are relatively close to the Earth, parallax will be significant and so a correction has to be made. These corrections are included in the altitude correction tables in the Nautical Almanac.

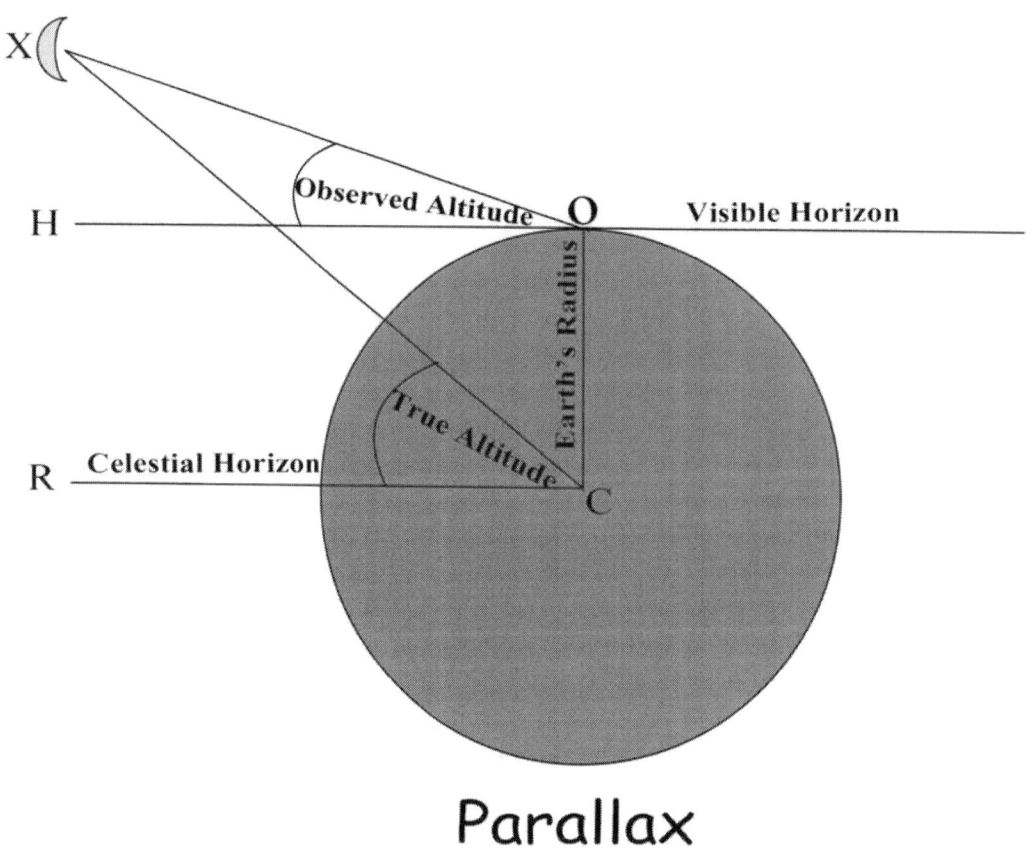

## Parallax

**Horizontal Parallax.** Parallax error is greatest when the celestial body is close to the horizon and decreases to zero as the altitude approaches 90°.

It is negligible except in the case of the Moon which is close to the Earth in comparison with the other celestial bodies. Because horizontal parallax is significant in the case of the Moon, a separate correction has to be applied. The hourly values of horizontal parallax for the Moon are listed in the daily pages of the Nautical Almanac.

# Chapter 4
# The Planets

There are 8 planets in our solar system and in order of their distances from the Sun, they are Mercury, Venus, Earth, Mars, Jupiter, Saturn, Uranus and Neptune.

The picture above includes Pluto which was classified as a planet until 2006; however, it is no longer considered to be one, even though it orbits the Sun in the same way that the planets do. Instead it is now classified as a dwarf planet and is relatively small compared to the existing planets, It is primarily made of rock and ice, and is only about 1/6 the mass of our Moon.

Looking from the surface of the Earth, five of the other planets: Mercury, Venus, Mars, Jupiter and Saturn can be seen by the naked eye but Neptune and Uranus can only be seen with the aid of a telescope or binoculars.

## The Inner Planets.

Mercury, Venus, Earth and Mars orbit relatively close to the Sun and are therefore known as the Inner Planets. They are also known as the Terrestrial Planets because they consist mostly of rocks and metals and have solid surfaces. They all have atmospheres which range from very thick on Venus to very thin on Mercury.

## The Outer Planets.

Jupiter, Saturn, Uranus and Neptune are known as the outer planets. They are, on average, ten times the mass of the Earth and are composed mostly of gases such as hydrogen and for these reasons, they are also known as the Gas Giants.

## The Dwarf Planets.

A dwarf planet is a celestial body that orbits the Sun but does not have sufficient mass to clear other objects out of its path in the way that the planets do.

There are 5 officially recognised dwarf planets in our solar system and these are Ceres, Pluto, Haumea, Makemake, and Eris. Ceres is located in the asteroid belt, is the largest of the dwarfs with Pluto, which is found in the outer solar system, taking second place.

## The Navigational Planets.

Of the seven planets in our solar system excluding Earth, only those that are sufficiently prominent to be observed with an ordinary sextant are considered to be 'navigational planets' and these are **Venus, Mars, Jupiter and Saturn.** Because this book is concerned with the application of astronomy to navigating on the surface of the Earth, we will focus our attention on just the navigational planets and their relationships with the Earth for the rest of this chapter. (Planet Earth is discussed in great detail in chapter 2).

# Venus.

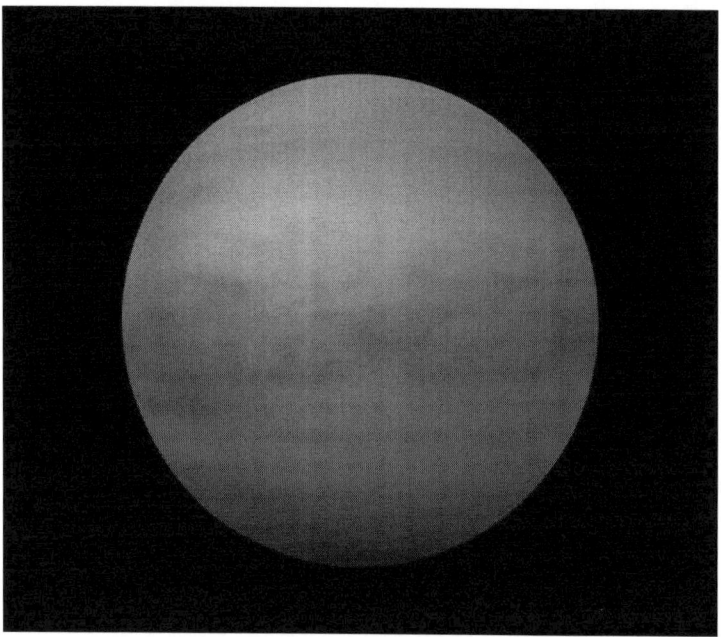

Venus is the second largest terrestrial planet, the Earth being the largest. It is named after the Roman goddess of love and beauty, and is sometimes referred to as the Earth's sister planet due their similar size and mass. Venus is the closest planet to the Earth; however, the distance between the two planets can vary by as much as 223 million km due to the fact that their elliptical orbits are not concentric. The closest that Venus can approach the Earth is 38 million km and at its farthest, it can be as far away as 261 million km.

After the Sun and the Moon, Venus is the third brightest celestial body in our night sky and for that reason, it is an important navigational planet.

### Venus – Evening Star or Morning Star?
Venus sometimes appears as an 'evening star' above the western horizon shortly after sunset and sometimes appears as a 'morning star' above the eastern horizon shortly before sunrise. In primitive times, people regarded the 'evening and morning stars' as two different heavenly bodies but in the sixth century BC, Pythagoras suggested that they might be one and the same body.

Ptolemy believed that the Earth was the centre of the Universe and that the Sun moved in a circular orbit around the Earth which was stationary. He also believed that Venus orbited the Sun as the Sun itself orbited the Earth and that this explained why Venus appeared from Earth as an 'evening star' during part of its orbit and as a 'morning star' during another part. The Ptolemaic hypothesis can be explained with the aid the diagram below:

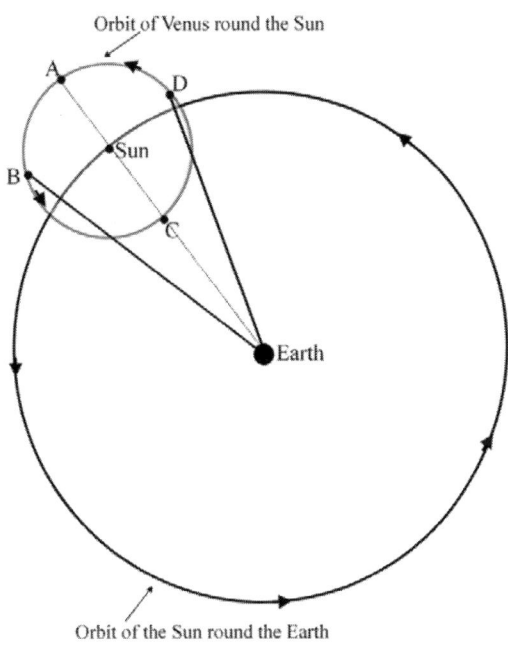

According to the Ptolemy, if Venus were at a point in its orbit somewhere on the semicircle ABC, say at B, then the line joining the Earth to Venus would be to the left of the line joining the Earth to the Sun. Therefore, looking from Earth, Venus would appear to be to the left of the Sun so, when the Sun set in the west, Venus would be seen for some time after sunset above the western horizon as an 'evening star'. In a similar way, if Venus were at a point somewhere on the semicircle ADC, say at D, it would appear from Earth to be to the right of the Sun and it would therefore appear above the eastern horizon just before sunrise as a 'morning star'.

We now know, thanks to the work of Copernicus, that both Earth and Venus orbit the Sun. According to the Copernican hypothesis, the reason that Venus sometimes appears as an 'evening star' and sometimes as a 'morning star' can be explained with the aid of another diagram:

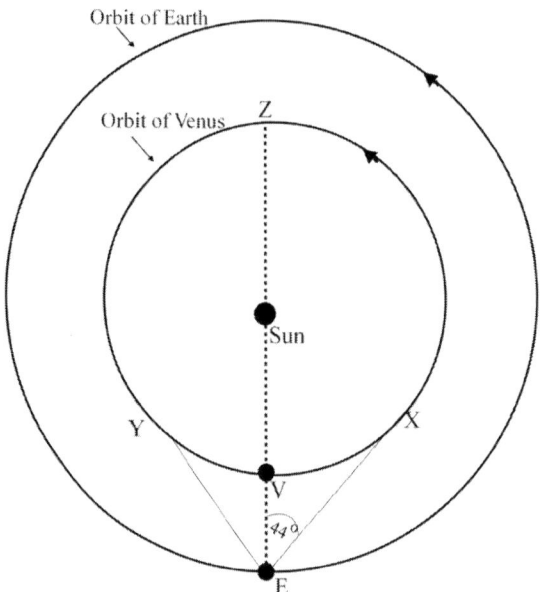

The average distance of Venus from the Sun is 108 million Km. while the average distance of Earth from the Sun is 150 million Km. Because the circumference of the orbit of Venus is smaller than that of the Earth, we can safely assume that Venus takes less time to complete its orbit than the Earth does. Whereas the Earth moves through 360° in 365.25 days, Venus completes its orbit in 225 days. For this reason Venus gains 0.6° per day on the Earth and overtakes it at intervals of approximately 600 days. From the diagram it will be seen that, after Venus has overtaken the Earth (that is after it has passed the point V) it becomes a 'morning star'. It increases its angular distance to the right of the Sun until it reaches 44° at point X. After passing X, the angular distance to the right of the Sun decreases until it reaches point Z, which is behind the Sun. After passing Z, it makes its appearance to the left of the Sun and is now an 'evening star'. As it approaches Y, its angular distance left of the Sun increases until it reaches 44° at point Y.

After passing Y on its way to V, its angular distance left of the Sun decreases until, at V, it is zero when it will pass either slightly above or below the Sun. On extremely rare occasions, it crosses in front of the Sun and this is known as a 'transit of the planet'.

To sum up, Venus overtakes the Earth at intervals of approximately 600 days. During approximately 300 of these days, it is a 'morning star' and for the other 300 days, it is an 'evening star'. The maximum angular distance right or left of the Sun, is roughly 44°.

**What about Mercury?** In a similar way, Mercury is also an 'evening star' and a 'morning star'. Mercury is the closest planet to the Sun and because it has the most elliptical orbit, its distance from the Sun ranges from 29 million Km. to 47 million Km. Because the circumference of its orbit is comparatively small, it gains 3° on the Earth per day and overtakes it on average every 120 days. For 60 of these days it will be a 'morning star' and for the other 60 it will be an 'evening star'. Its maximum angular distance left or right of the Sun is roughly 24°.

So Venus and Mercury are both 'morning stars' and 'evening stars' but it is quite easy to tell them apart for the following reasons: Venus usually appears much higher in the sky than Mercury and is far brighter.

## Mars, 'The Red Planet'.

Mars is named after the Roman god of war. It is the fourth planet from the Sun and is one of the terrestrial planets. Its surface is composed mostly of iron oxide which gives it a red colour and this makes it very easy to locate in the sky so adding to its value as a navigational planet.

The mean distance of Mars from the Sun is 227.9 km (141.71 million miles) which is 78.31 million km greater than the mean distance from the Earth to the Sun. It is about half the size of Earth in diameter and it rotates about its axis in 24 Earth hours, 37 minutes, 23 seconds.

Like the other planets, Mars orbits the Sun in an eccentric, elliptical orbit so the distance to Mars from Earth is constantly changing. The closest the planets could come together is 54.6 million km (33.9 million miles) and the farthest apart they can be is about 401 million km (250 million miles).

The orbital path of Mars is more complex than those of Mercury and Venus and we can best understand it by considering its motion relevant to the distant stars. To do this, we must study the effect of the motions of both the Earth and Mars around the Sun.

Firstly, the circumference of Mars orbit is 1.6 times greater than that of the Earth and with an orbital speed of 24 km/s, it takes 686.98 Earth days to complete each orbit while Earth travels at 30 km/s taking 365.25 days. Thus, the Mars' year is 1.8 times as long as the Earth year.

Secondly, the orbit of the Earth is closer to the Sun than that of Mars, and this puts Earth on the inside track, so to speak. Taking this into account along with Earth's greater speed, it is easy to see that the Earth will overtake Mars from time to time; in fact it laps Mars every 26 months.

The next diagram represents the orbits of the Earth and Mars around the Sun.

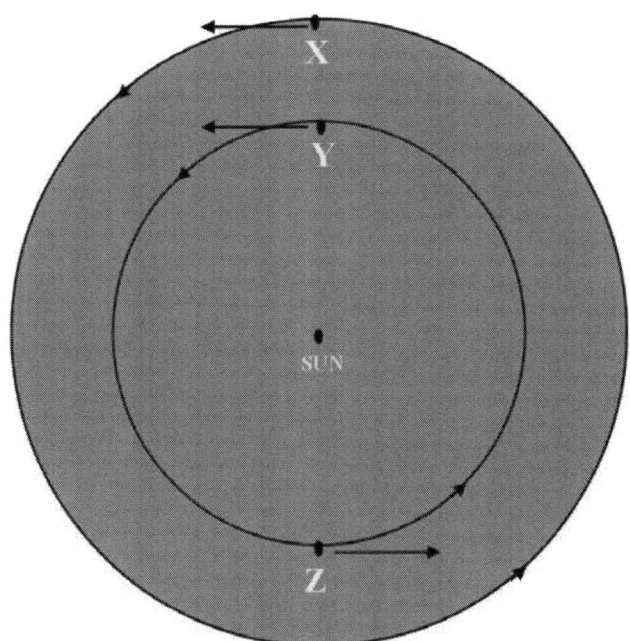

## Mars Retrograde Motion

When Mars is at point X and Earth is at point Y, Mars and Earth are at their closest and although they are both moving from right to left of the diagram, because Mars is travelling 6 km/s slower, it will appear, to an observer on the Earth, to be moving left to right. This is called 'retrograde motion'. However, if the Earth is at point Z, it will appear that Mars is moving in the opposite direction, that is right to left. The speed at which the two planets will be moving in opposite directions will be equal to their combined orbital speeds; i.e. 54 km/s. This is called prograde motion'.

These changes in the apparent motion of Mars from retrograde to prograde and vice versa are not sudden changes. Before a change in direction, the planet seems to slow down and then pause for about a week before starting to move in the new direction.

To summarise, Mars will sometimes appear to be moving from left to right with respect to the background stars, sometimes it will seem to move in the opposite direction and in between these changes in, it will appear to pause.

# Jupiter

Aptly named after the Roman 'King of the Gods', Jupiter has 317 times the mass of the Earth and two and a half times the mass of all the other planets in our solar system combined. It is the fifth planet out from the Sun and is one of the outer planets. It is also one of the 'gas giants' being largely composed of gases, primarily hydrogen. The distance of Jupiter from the Sun is constantly changing because, like the other planets, its orbit is elliptical; the mean distance is 778,300,000 km. It travels a distance of 4,887,600,000 km at a speed of 13 km/s to complete each orbit taking 4,332.82 Earth days or 11.85 Earth years to do so.

Jupiter appears sometimes an 'evening star' and sometimes as a 'morning star' but not for the same reasons as Venus. Jupiter will disappear behind the Sun when it is in conjunction with the Earth, that is when the two planets are on opposite sides of the Sun. As it approaches conjunction, Jupiter appears as an 'evening star' near the western horizon at twilight. When it emerges from behind the Sun, it appears as a 'morning star' near the eastern horizon. This process is repeated at intervals of approximately 13 months when Jupiter and Earth approach conjunction.

Just as the movement of Mars appears to change between retrograde and prograde, so does that of Jupiter. The topic of retrograde motion was dealt with in our discussion of Mars so we do not need to go into it in any great detail when discussing Jupiter.

Jupiter moves across the sky in a very predictable pattern, but every now and then it reverses direction in the sky, making a tiny loop against the background stars – this is Jupiter in retrograde.
The following diagram shows that, as Jupiter is overtaken by the Earth, its apparent motion across the sky appears to describe a loop as its direction changes from prograde to retrograde and then back to prograde again.

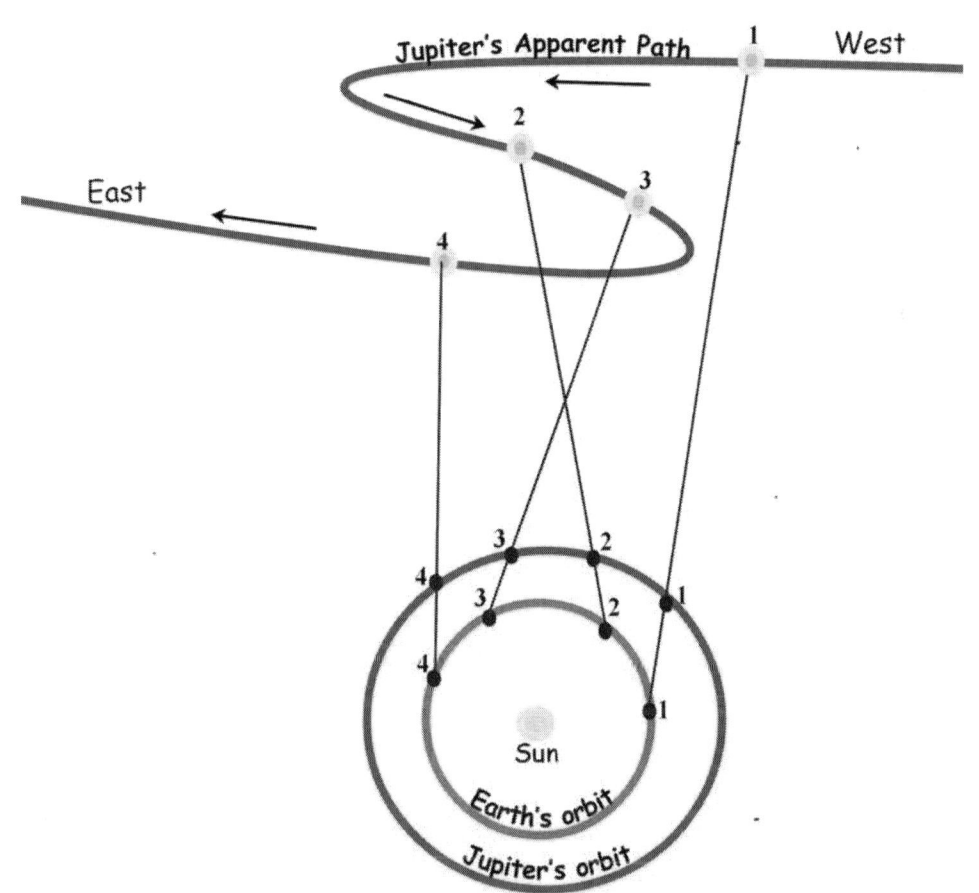

**Jupiter's Retrograde Loop**

At position 1, it appears to be moving from west to east in prograde motion. At positions 2 and 3, its direction appears to have changed from prograde to retrograde so that it is now moving from east to west. At position 4, it appears to have resumed prograde motion as it moves from west to east again.

**Note.** Sky maps can be very confusing because they are not drawn in the conventional way with east on the right and west on the left. They are drawn as they would appear in the sky if we were lying down with our legs pointing to the south and looking upwards so that east would be on our left and west on our right.

Jupiter's retrograde periods last for 4 months and are then followed by periods of nine months of prograde motion before going retrograde again. So the time from the beginning of one retrograde movement to the beginning of the next is approximately 13 months.

The relatively slow movement of Jupiter across the sky makes it very easy for the navigator to locate. It appears to move from one constellation to another every 13 months describing its retrograde loop as it pauses in each one.

In January 2016, Jupiter will pause in Leo when it goes into retrograde motion and thirteen months later, in February 2017 it will be in Virgo when it begins the sequence again. After that Jupiter is retrograde in Libra in March 2018 and so on as it follows its path through Scorpio, Sagittarius, Capricorn, Aquarius, Pisces, Aries, Taurus, Gemini, Cancer and back to Leo again.

Jupiter's predictable path across the sky together with the fact that it is the fourth brightest celestial body in the sky explains why it is such an important navigational planet.

## Saturn.

Named after the Roman god of agriculture, Saturn has similar characteristics to the other planets that we have discussed. It is the sixth planet from the Sun and is the second largest object in the solar system following Jupiter. Like Jupiter it is one of the gas giants and is composed mainly of hydrogen. It is the fifth brightest object in the solar system after the Sun, Moon, Venus and Jupiter and is classified as a navigational planet.

Like Jupiter, Saturn disappears behind the Sun when it is in conjunction with the Earth; however, unlike Jupiter, it does not become a 'morning or evening star' as it is not visible for about two weeks either side of conjunction.

The distance of Saturn from the Sun is constantly changing because, like the other planets, its orbit is elliptical; the mean distance is 1.4 billion km. It travels a distance of 8.4 billion km at a speed of 9.69 km/s and takes 354 Earth months (29.5 Earth years) to complete each orbit.

Like Mars and Jupiter, Saturn's apparent motion changes between prograde and retrograde. Each of its retrograde loops lasts for an average of 140 days and the average period between each phase is an average of approximately 12.4 months.

Like Jupiter, Saturn follows a predictable path across the sky which helps us to locate it.  From 2015 to 2022 it moves slowly in an easterly direction through the constellations of the southern hemisphere describing its retrograde loops at intervals of roughly 12 Earth months as it goes.  In late 2015 and throughout 2016 it passes to the north of Scorpius reaching the north eastern boundary of Sagittarius in 2017.  It continues its journey through Sagittarius during 2018, 2019 and 2020 until it reaches Capricorn in 2021.  It spends the next three years moving slowly through Capricorn after which it continues its path until it returns to Scorpius in 2043 and begins the cycle all over again.

So Saturn is an easy planet for a navigator to find; not only is it is the fifth brightest object in the sky, we always know just where to look for it as it slowly follows the path described above.

# Chapter 5
# Stars and Constellations
(Route Map To The Stars)

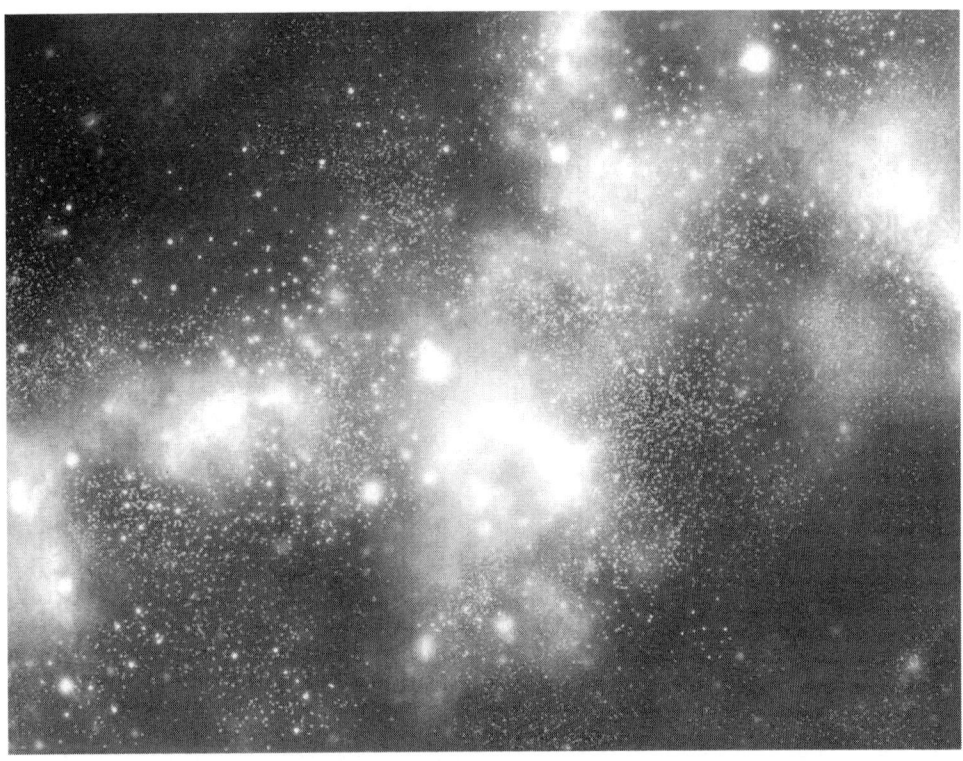

Of all the stars that are visible in the sky, only 59 are considered to be bright enough to be seen at nautical twilight when star sights are usually taken; these are known as the **Navigational stars**. What we need to know in astro navigation therefore, is how to locate the navigational stars in the sky and that is the focus of this chapter. As a general rule, only constellations that contain navigational stars are dealt with here; however, some constellations that do not contain navigational stars themselves but provide pointers to other constellations that do, are also included as are those constellations which have special interest. Throughout this chapter, the names of the navigational stars have been emboldened as an aid to identification.

The stars are at such an immense distance from the Earth that the movement of their relative positions in the sky is very slow and very small and can be discounted without any great loss of accuracy; we can assume therefore, that they are in fixed positions in the celestial sphere. The sun, moon and planets however, are relatively close to the Earth and have varying apparent movements and speeds about the Earth. For this reason, the positions of these bodies have to be accurately calculated if they are used for position fixing.

The focus of this chapter is to provide a 'route map to the stars' that is, to demonstrate ways of locating stars, specifically navigational stars; it is not intended to be an astrological tour of the constellations of the zodiac. However, the order of the signs of the zodiac can be very useful in astro navigation because it tells us the relative position of any zodiac constellation with respect to the others. For example, we know that Cancer's position in the zodiac falls between Leo and Gemini and as the following diagram demonstrates, Cancer can easily be found nestling on the ecliptic between those constellations with Gemini lying to the west and Leo to the east From this, we can see that knowing the position of one zodiac constellation in the sky can help us to locate those adjacent to it along the path of the ecliptic

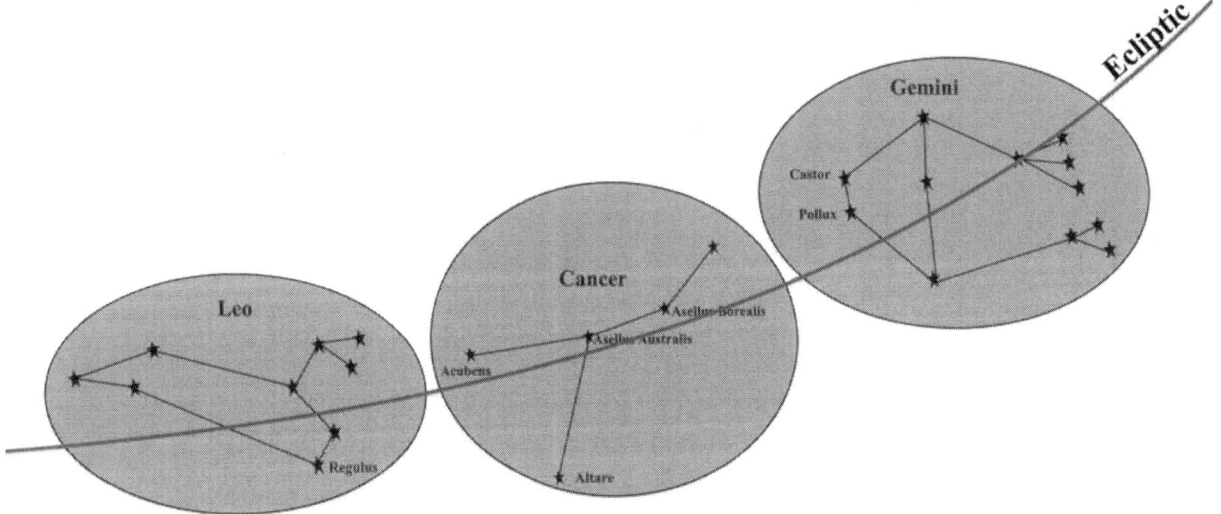

Star maps can be very confusing because they are not drawn in the conventional way with east on the right and west on the left. They are drawn as they would appear in the sky if we were lying down with our legs pointing to the south and looking upwards so that east would be on our left and west on our right.

The easiest way to locate the stars in which we are interested is to locate the constellations to which they belong.  One method of doing this is to establish reference lines in known constellations and from these to memorize the directions in which other constellations lie in the celestial sphere.

We should remember that, as the Earth spins on its axis, the positions and orientation of the stars and constellations in the sky will change.  The Earth spins about its axis from west to east and as it does so, the stars in the sky will appear to rotate around the Pole Star from east to west.  Stars that are within the arctic and Antarctic circles and which are therefore close to the poles will be circumpolar which means that they are always above the horizon but others will rise and set in the same way that the Sun and moon do.

Our time system is based on the Mean Solar Day which, as explained in chapter 7, is 24 hours (the time taken for the Mean Sun to make one complete circuit of the Earth).  However, we learned in chapter 2 that the stars will complete their circuit of the Earth in 23 hours, 56 minutes and 4 seconds so, according to our clocks, the stars will rise and set 3 minutes and 56 seconds earlier each day (usually rounded off to 4 minutes).  This means that except for those that are circumpolar, there will be long periods between the times that the navigational stars are above the horizon at the right times to be available for navigation purposes.
The optimum time for taking star sights is during nautical twilight when the sky is light enough for the horizon to be seen yet dark enough for the stars to be seen.

Of course, it is not sufficient to know only the direction in which a constellation lies with respect to another constellation; it is also necessary to know the distance between them.  This presents us with a problem: whereas we measure distances on the surface of the Earth in miles or kilometres, in space these units of measurement have no significance.  Instead, to measure distances between objects in space, we use degrees or angular distance as we say.  The following 'rule of thumb' provides a crude but surprisingly accurate method of estimating the angular distance between two bodies:
- The span of the hand when held at arm's length from the body will subtend an angle of roughly 20° at the eye of the observer.

- The width of the palm of the hand will subtend an angle of approximately 10°.
- The width of the first finger subtends an angle of approximately 1°.

For example, the angular distance between Star Duhbe in the Big Dipper and the Pole Star is 28° or roughly three palm widths.

Although this route map can help us to find the relative position of a particular constellation with respect to other constellations, it will not necessarily indicate whether or not it will be visible above the horizon. That will of course depend on its times of rising and setting. (Chapter 8 gives an exposition of the 'Where to Look' Method which enables us to calculate when a celestial body will be above the horizon).

## Ursa Major, The Great Bear (also known as The Big Dipper or the Plough).

The best known and easily recognizable constellation in the northern hemisphere is the constellation Ursa Major which is also known by various names such as the Great Bear, the Big Dipper and the Plough.

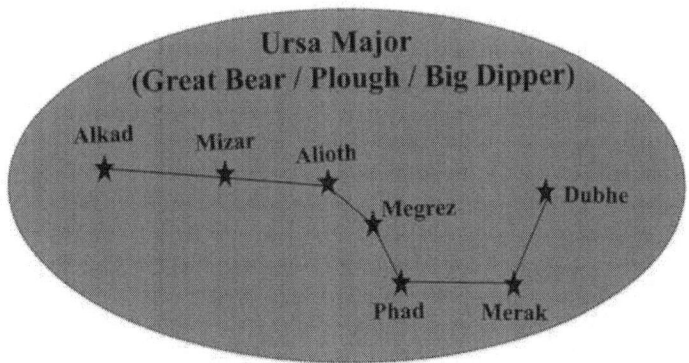

This circumpolar constellation is visible all year round in the northern hemisphere and in northern regions of the southern hemisphere; it contains 3 navigational stars named **Dubhe**, **Alioth** and **Alkaid**.

## Ursa Minor, The Little Bear (also known as the Little Dipper).

Ursa Minor is also a circumpolar constellation which never sets in the northern hemisphere and which can be seen as far south as 10°S.

**Polaris.** Although Polaris (otherwise known by various names including Pole Star, North Star, Lodestar and the Guiding Star) is the brightest star in Ursa Minor, it is only the 45th brightest star in the sky and is not listed as a navigational star in the nautical almanac issued by the United Kingdom Hydrographic Office. However, it has always played an important role in navigation; not only because it indicates the direction of north but also because it can be used for position fixing particularly in the polar-regions and for this reason, it is listed in the American Practical Navigator as a navigational star. Ursa Minor contains one other navigational star, Beta Ursae Minoris which is otherwise known by its traditional name of **Kochab**.

## Finding the Pole Star.
Ursa Major contains a reference line known as the line of pointers. The line joining Merak to Dubhe, when extended will point to Polaris in the constellation Ursa Minor as illustrated in the following diagram.

© Copyright Jack Case 2015

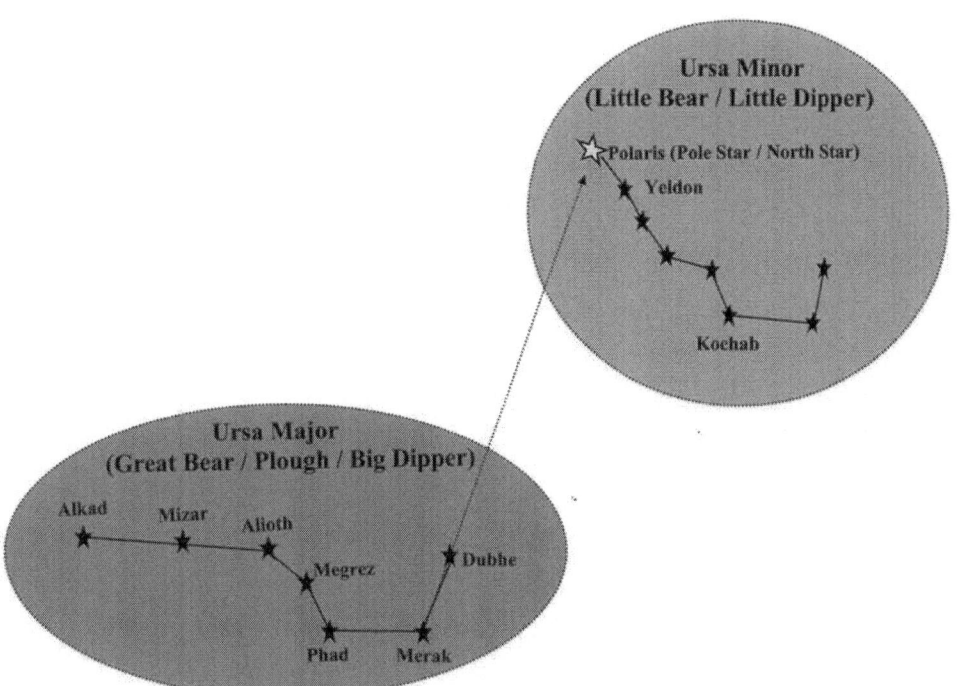

Ursa Major and Ursa Minor are associated with several mythological stories. In one such story, Ursa Major is associated with Callisto, a beautiful young nymph with whom Zeus had fallen in love. When Hera, the wife of Zeus found out, she turned Callisto into a bear. Ursa Minor is identified with Arcas, the son of Callisto who did not recognize his mother when she was in bear form and attempted to kill her until Zeus stepped in and sent them both into the sky.

In yet another story, the two nymphs who had nursed Zeus as an infant were sent into the sky by him to form constellations. The nymph Adrasteia became the constellation Ursa Major while the other, Ida became Ursa Minor.

Later stories identified Ursa Minor with Hesperides, one of the daughters of Atlas who guarded Hera's temple which had an orchard where apples that gave immortality grew. In these stories, Arcas became identified with the constellation Boötes.

Leo, The Lion.

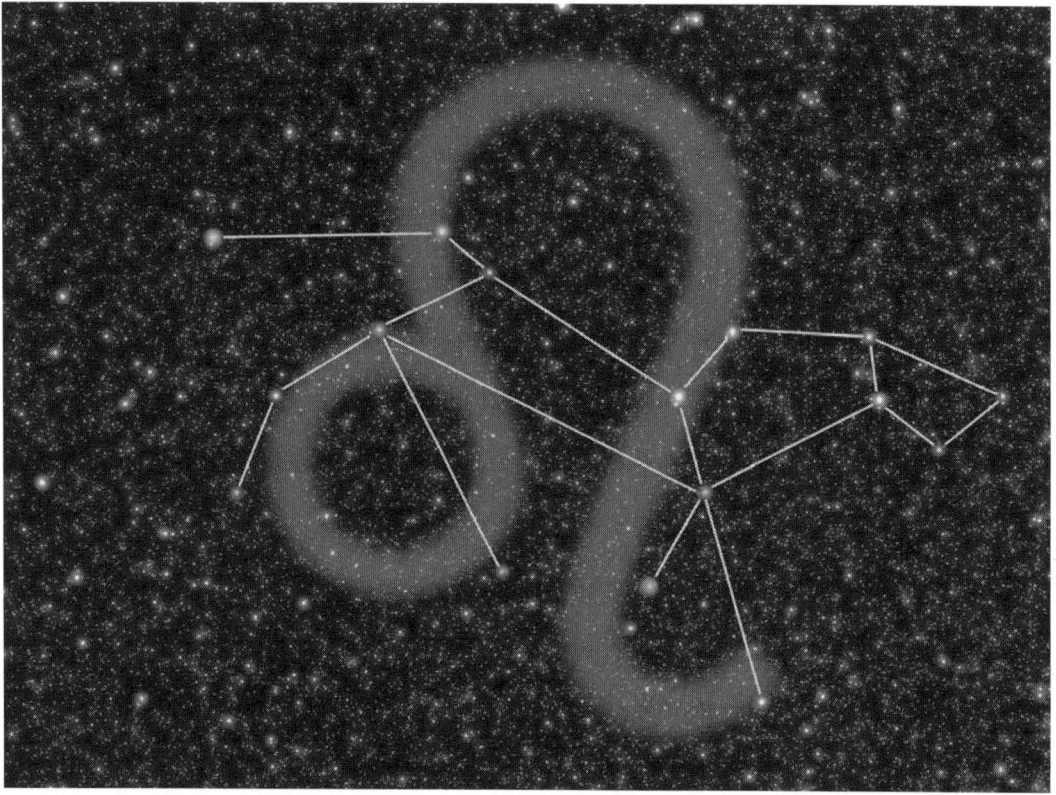

Leo is one of the largest constellations in the sky and is visible throughout the Northern Hemisphere during the winter and spring months and in the northern regions of the southern hemisphere during the summer and autumn months.

The name Leo means Lion in Latin and the constellation, which is depicted as a crouching lion, is associated in Greek mythology with the lion of Nemea which was killed by Heracles as one of his twelve labours.

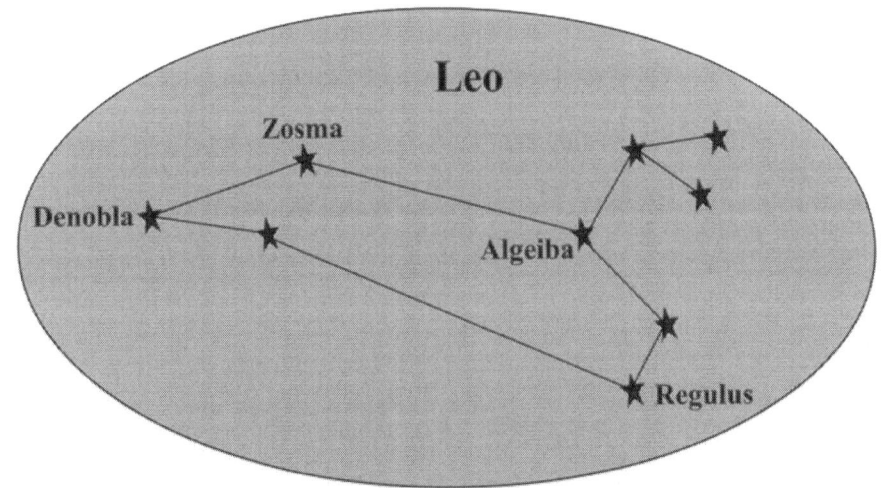

Leo is home to two navigational stars, **Denobola** and **Regulus** which are shown in the diagram above. Regulus, the brightest star in the constellation, is said to mark the lion's heart with Denobola marking the tip of its tail. From a navigation perspective, these stars are best seen during nautical twilight during the month of April.

**How to find Leo.**

When the line of pointers in Ursa Major is produced in the opposite direction to the Pole Star, that is from Dubhe to Merak, it will point to Leo as shown in the diagram below.

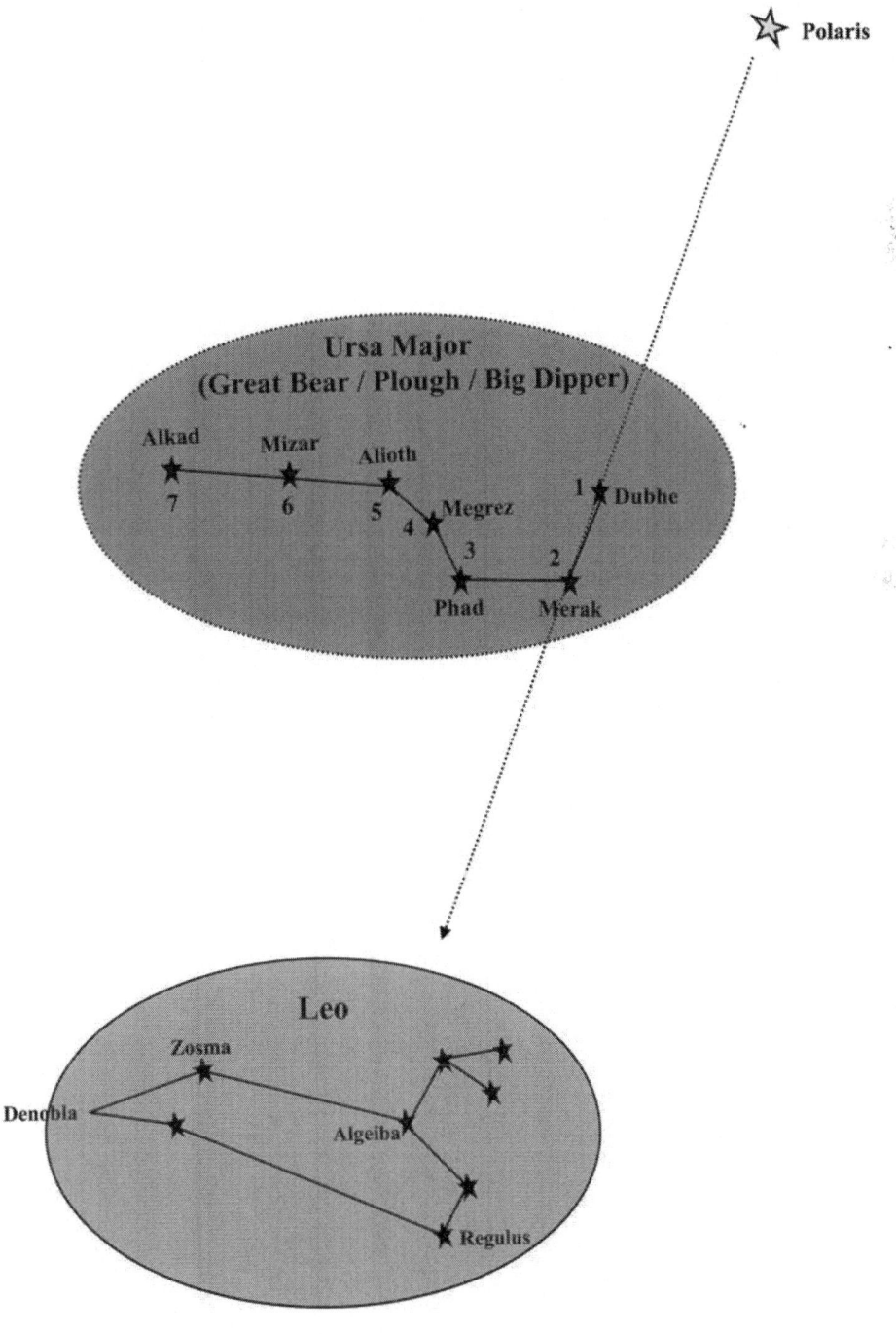

# Gemini, The Heavenly Twins.

The name Gemini means Twins in Latin and for this reason, it has the alternative name the 'Heavenly Twins'. The constellation is associated in, Greek mythology, with the twins Castor and Polydeuces. Pollux and Castor, the brightest stars in the constellation are said to form the eyes of the twins. (We use the Latin name Pollux instead of the Greek name Polydeuces).

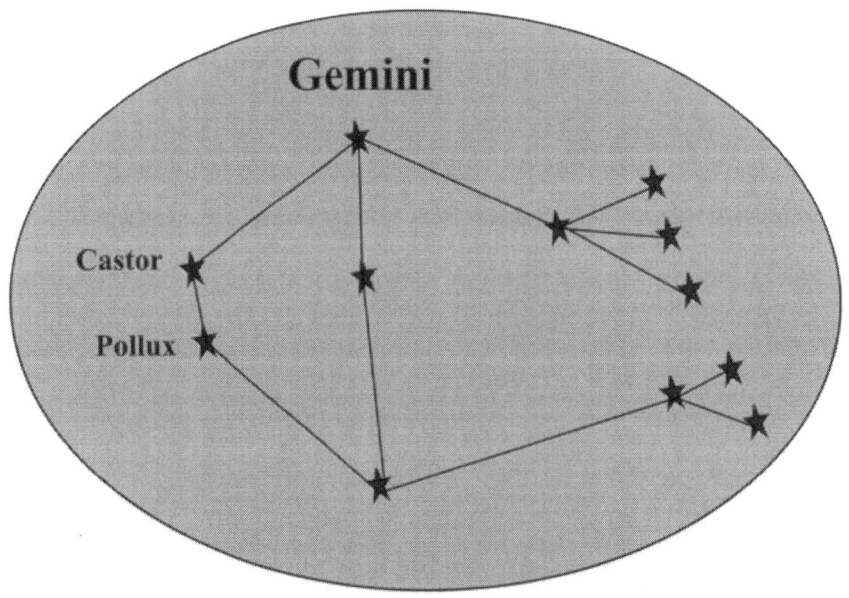

Gemini is in the northern hemisphere and is visible between 90°N and 60°S. **Pollux** is a navigational star and is best seen for navigation purposes during nautical twilight in February.

**Finding Gemini.** As shown in the diagram below, a line running from Megrez to Merak in Ursa Major leads to the constellation Gemini.

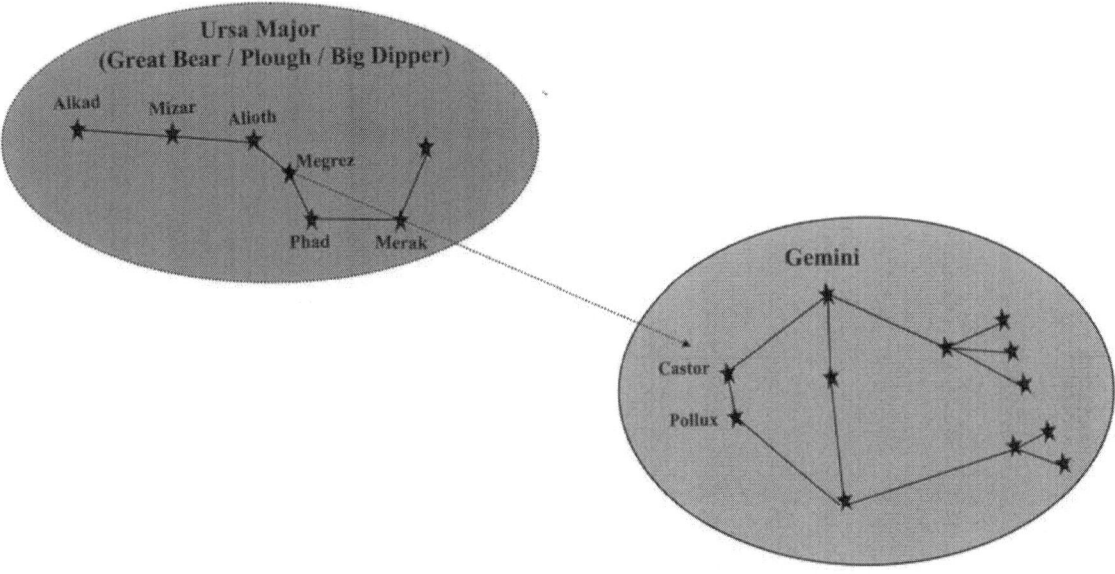

**Cassiopeia.**

**How to find Cassiopeia.** Cassiopeia can be located along a line of reference from the Pole Star at an angle of 135° to the line of pointers in Ursa Major, as the diagram below shows. As Ursa Major revolves around the Pole Star, so do the five stars of Cassiopeia but Segin always keeps its position 135° from the line of pointers. The angular distance of star Segin from the Pole Star is 30° or roughly one and a half hand-spans.

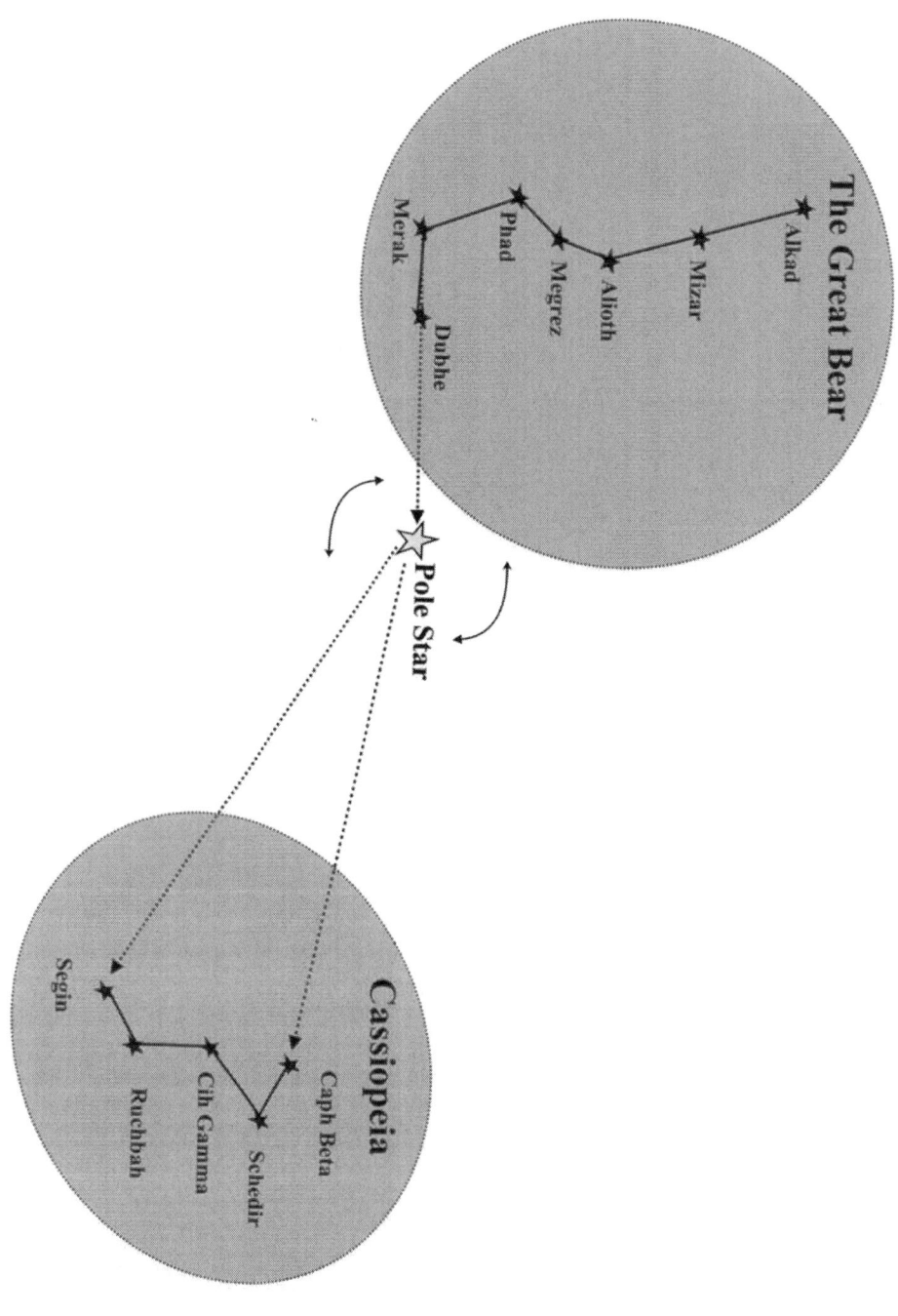

Cassiopeia is quite an easy constellation to find because of its 'W' shape which sometimes hangs upside down as it circles the pole. Its brightest star is Alpha Cassiopeia otherwise known as **Schedir** which is a navigational star.

The constellation Cassiopeia is associated with Queen Cassiopeia in Greek mythology. In punishment for her vanity, she was made to sacrifice her daughter Andromeda and as further punishment, she was sent into the sky to circle the North Pole forever.

The five stars in the constellation Cassiopeia appear in the celestial sphere in the shape of the letter W and can be observed throughout the northern hemisphere and down to 20°S.

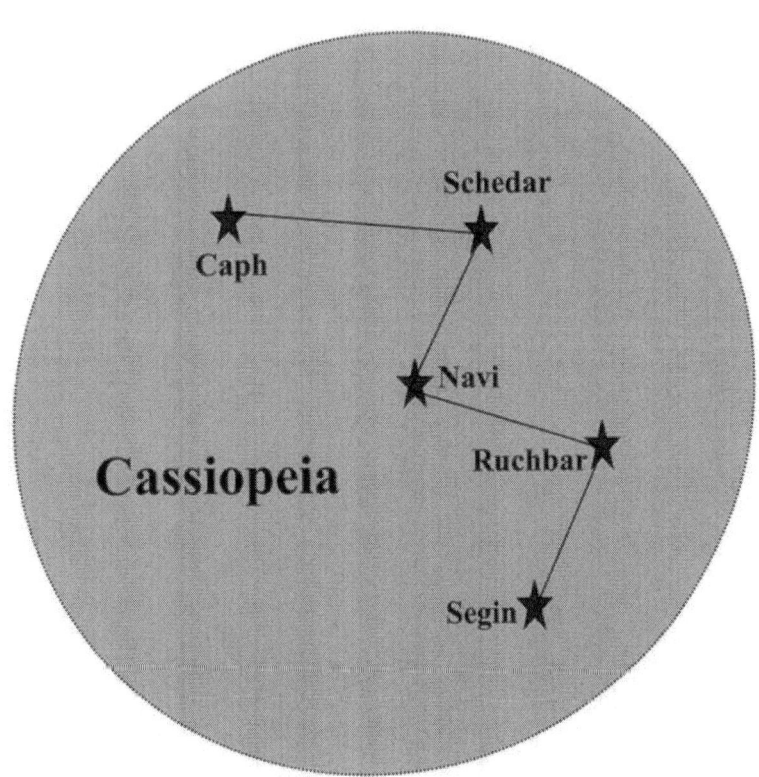

Perseus.

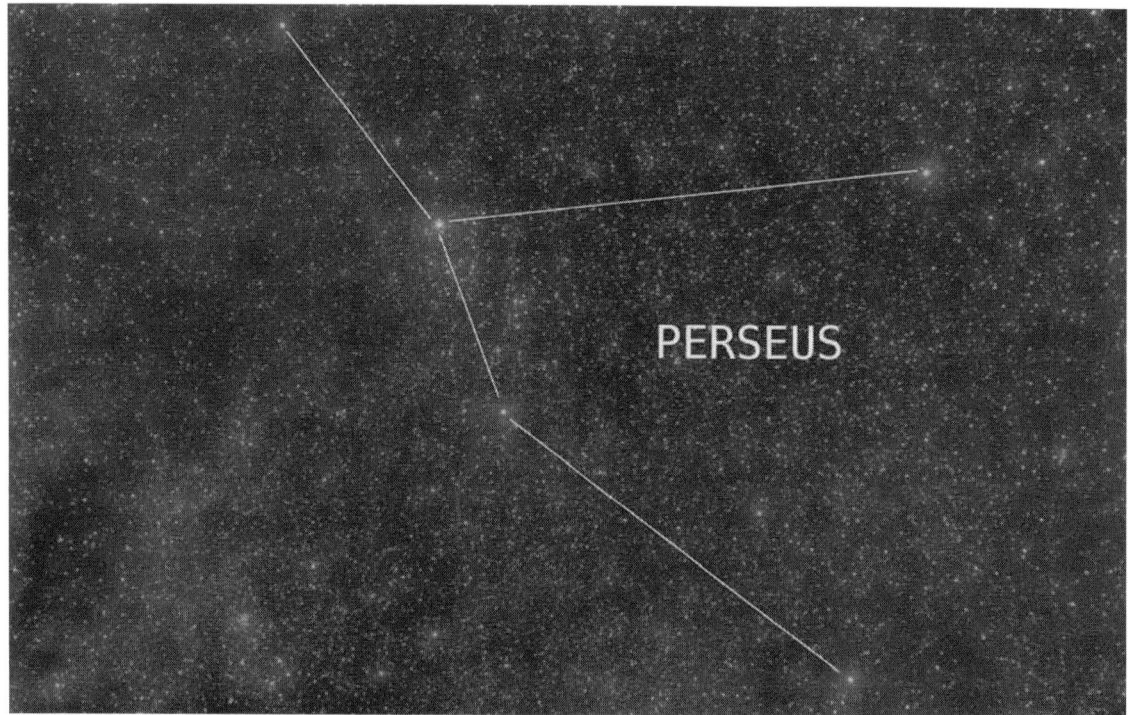

The constellation Perseus can be seen in the northern hemisphere during the winter months and in the northern areas of the southern hemisphere north of 35°S during the summer.

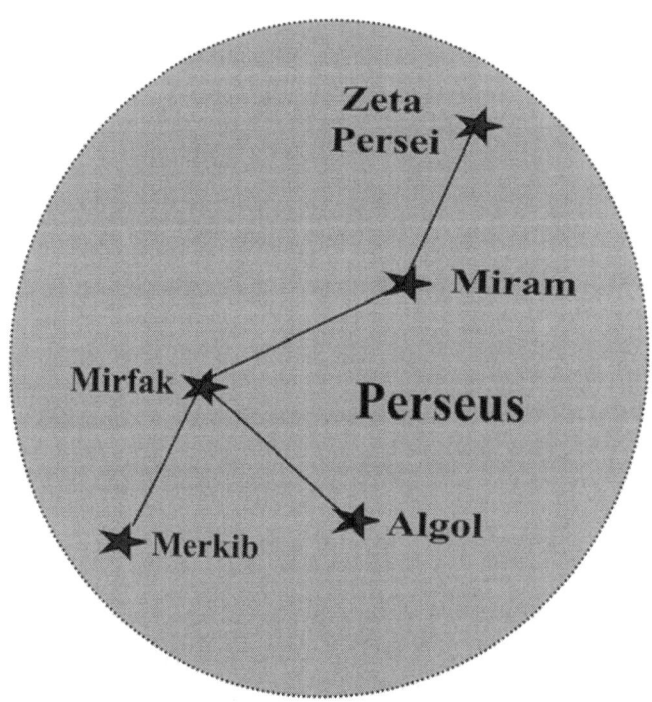

**Finding Perseus.** If a line is drawn from Navi to Ruchbah in Cassiopeia, it will point almost directly towards the star Mirfak of the constellation Perseus at about one hand-span as shown in the diagram below.

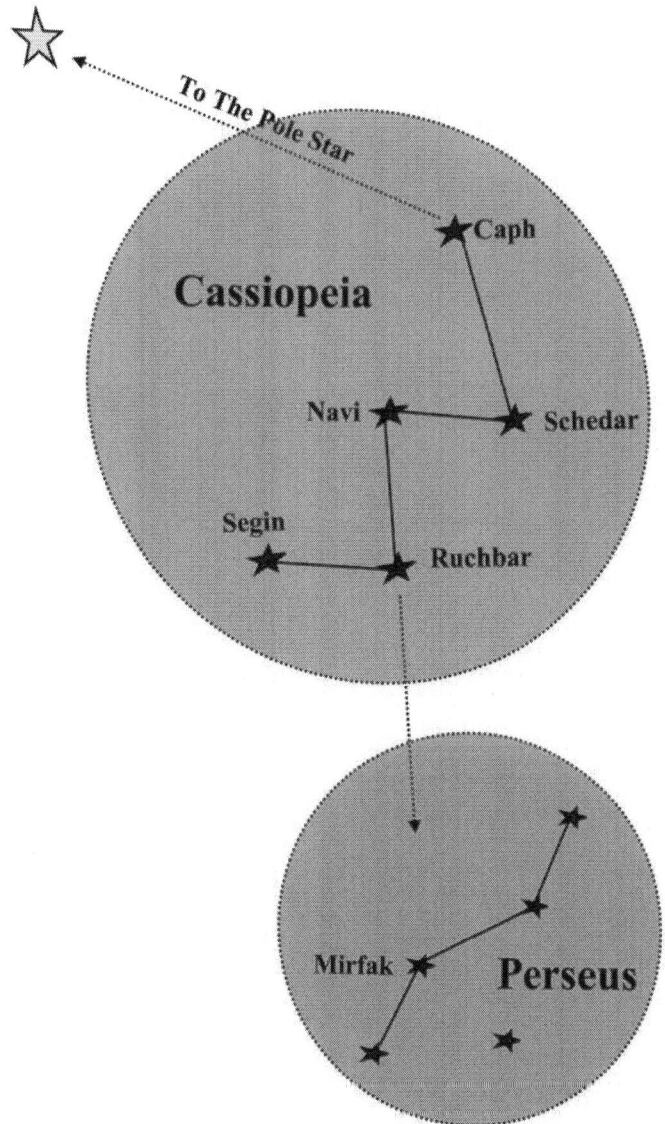

This constellation is associated with the Greek mythological hero Perseus who, on the orders of King Polydectes, slayed the Gorgon Medusa who had the power to turn people to stone. Polydectes had hoped that Perseus would not return and when he did, he became hostile. Perseus was so angered by this that he took out the head of Medusa and turned Polydectes to stone.

The wife of Perseus was called Andromeda and the constellation that represents her lies side by side with the one that represents him.

Even though Perseus' stars are bright relative to other constellations, **Mirfak** is its only navigational star and this is best seen for navigation purposes during nautical twilight in December.

## Andromeda. The Chained Maiden.

Andromeda is a constellation in the northern hemisphere and is visible between latitudes 90°N and 60°S.

The brightest star in Andromeda is Alpheratz which is also included in the constellation Pegasus. **Alpheratz** is a navigational star which is best seen during nautical twilight in the month of November.

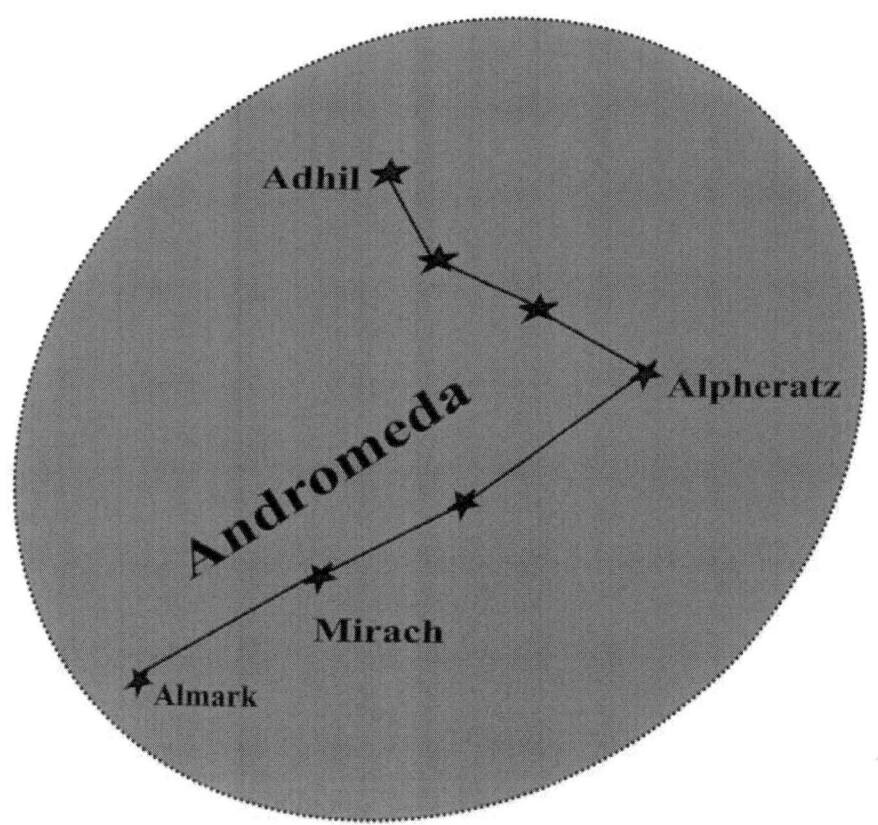

The constellation Andromeda is named after Andromeda, the wife of Perseus in Greek Mythology. It is sometimes called the 'Chained Maiden' because according to the legend, Andromeda was rescued by Perseus who found her chained to a rock and left as a sacrifice to the monster Cetus.

**How to find Andromeda.** The next diagram shows that, if a line from the Pole Star to Segin in Cassiopeia is extended by about one hand-span, it will point to the star Almark of Andromeda and that a line from the Pole Star through the star Caph Beta of Cassiopeia will point to Alpharatz in Andromeda.

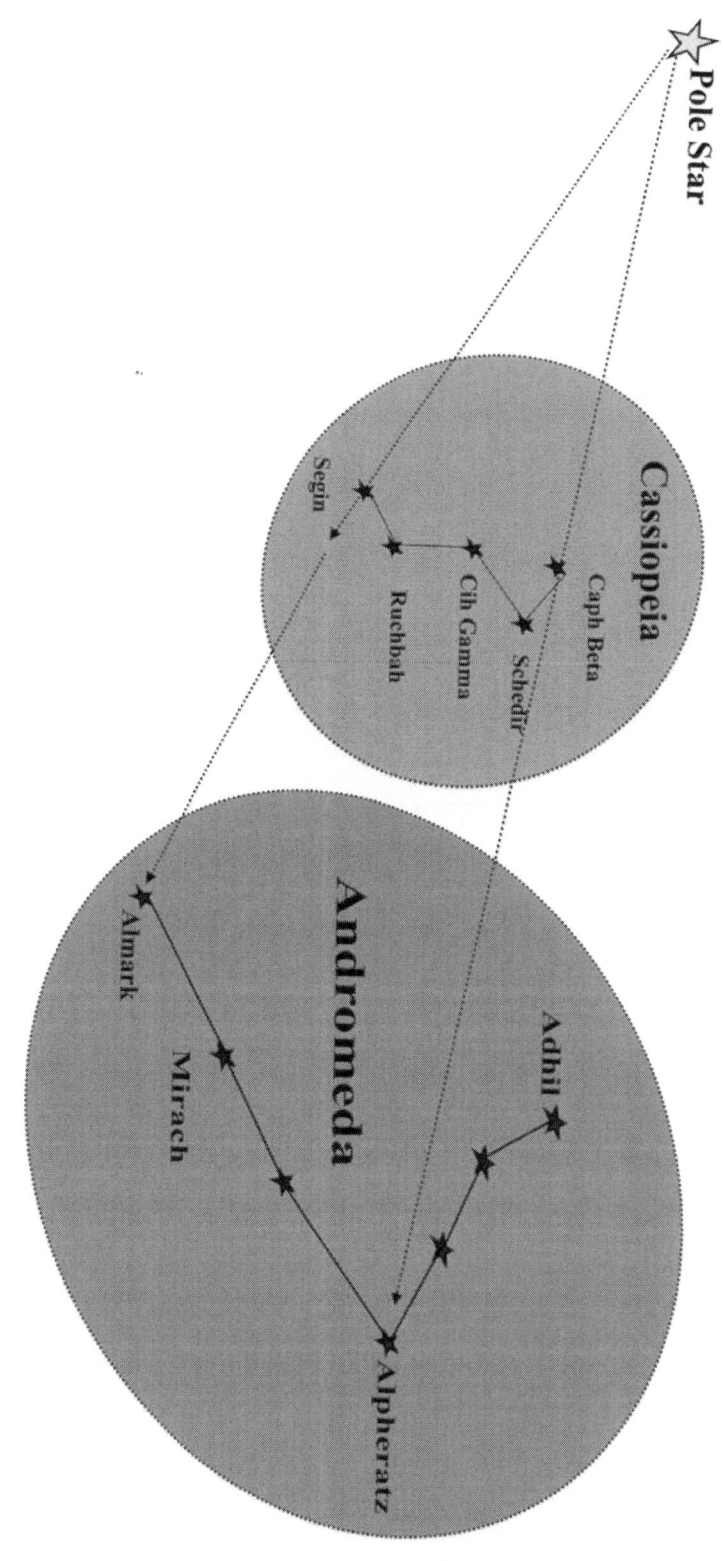

# Pegasus

As shown in the following diagram, once we have found Andromeda, we will also have found the constellation Pegasus because the star Alpheratz in Andromeda is also included in what astronomers call the 'The Great Square of Pegasus'.

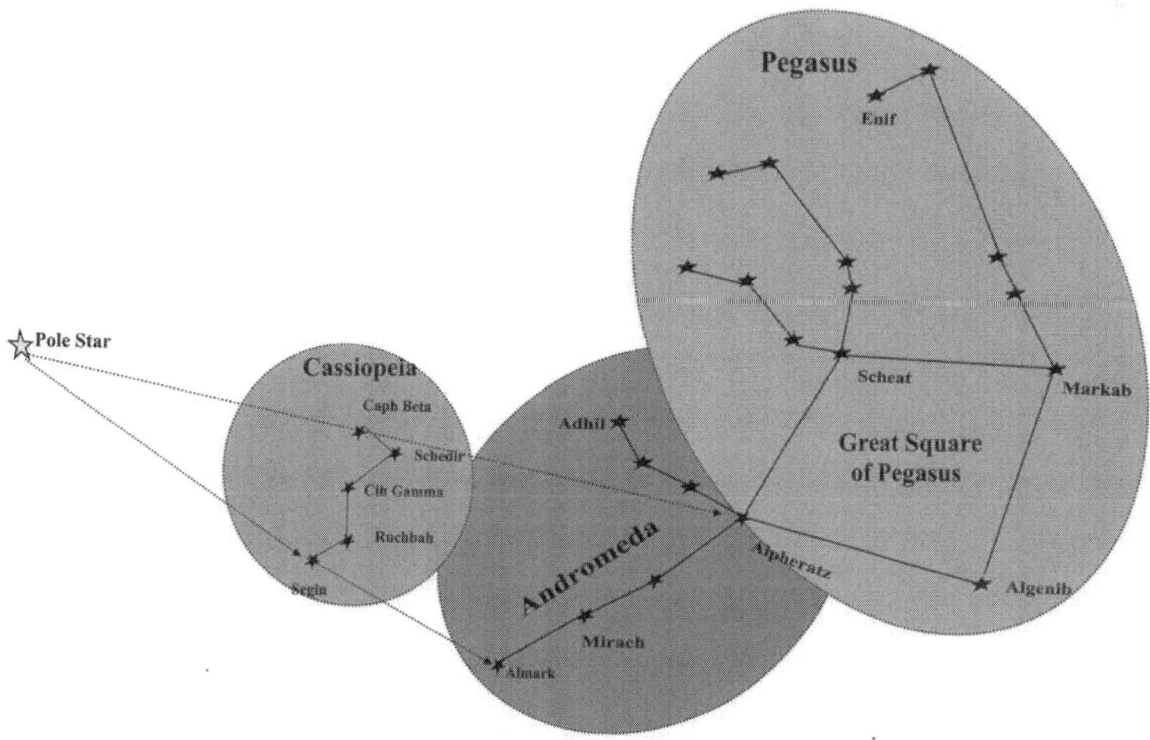

Pegasus, which is the 7th largest constellation in the sky, is in the northern hemisphere and can be seen from 90°N to 60°S. It has two navigational stars, **Enif** and **Markab** which are best seen during nautical twilight in the month of October.

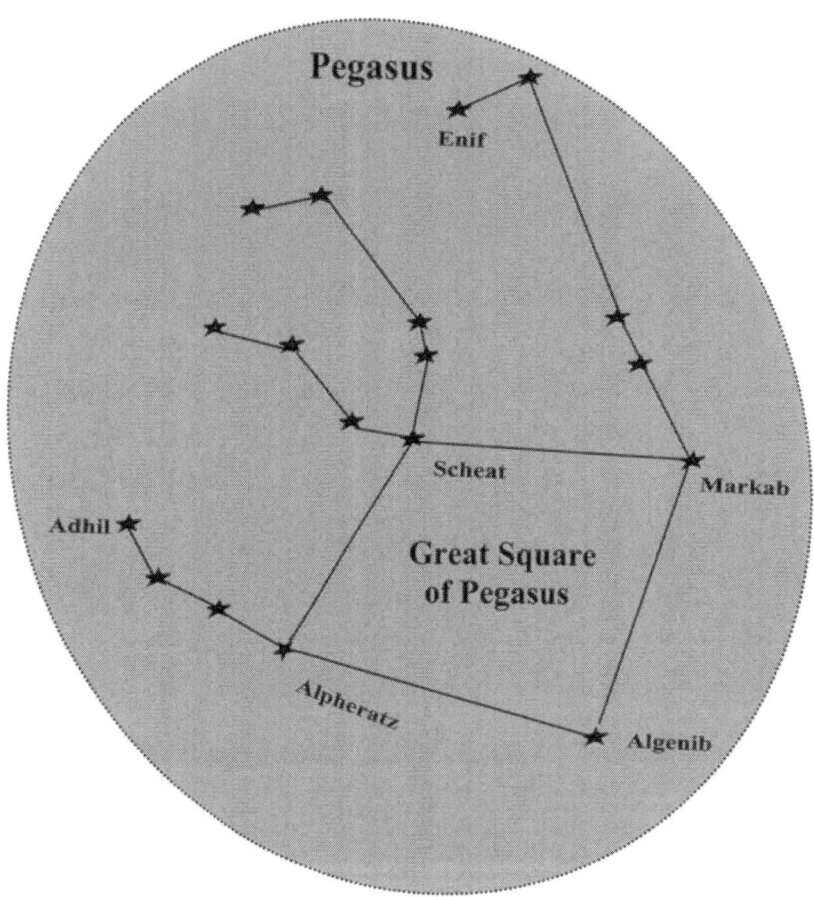

As briefly mentioned above, the 'Square of Pegasus' is a large asterism which is said to mark the body of the winged horse. This asterism is formed by the stars Scheat, Markab, Algenib and Alpheratz (which is also in Andromeda).
The brightest star in Pegasus is Enif which is said to mark the horse's nose.

In Greek mythology, the winged horse Pegasus is said to have leaped from the body of the Gorgon Medusa after she had been slain by Perseus. The hero Bellerophon tamed the winged horse and tried to ride it to Olympus. However, Bellerophon fell from Pegasus but the horse made it to Olympus where it was kept by Zeus to carry his thunder and lightning.

## Lyra, Cygnus and Aquila

The diagram below shows the constellations Lyra, Cygnus and Aquila.
If we join the star Meral to a point midway between stars Alioth and Dubhe in Ursa Major and then extend this line, it will point to Vega, the brightest star in the constellation Lyra.

**The Summer Triangle.**  The stars Deneb in the constellation Cygnus, Altair in Aquilla and Vega in Lyra form an astronomical asterism which is known as the 'Summer Triangle'. The diagram above shows how the triangle is formed by imaginary lines drawn between those stars.

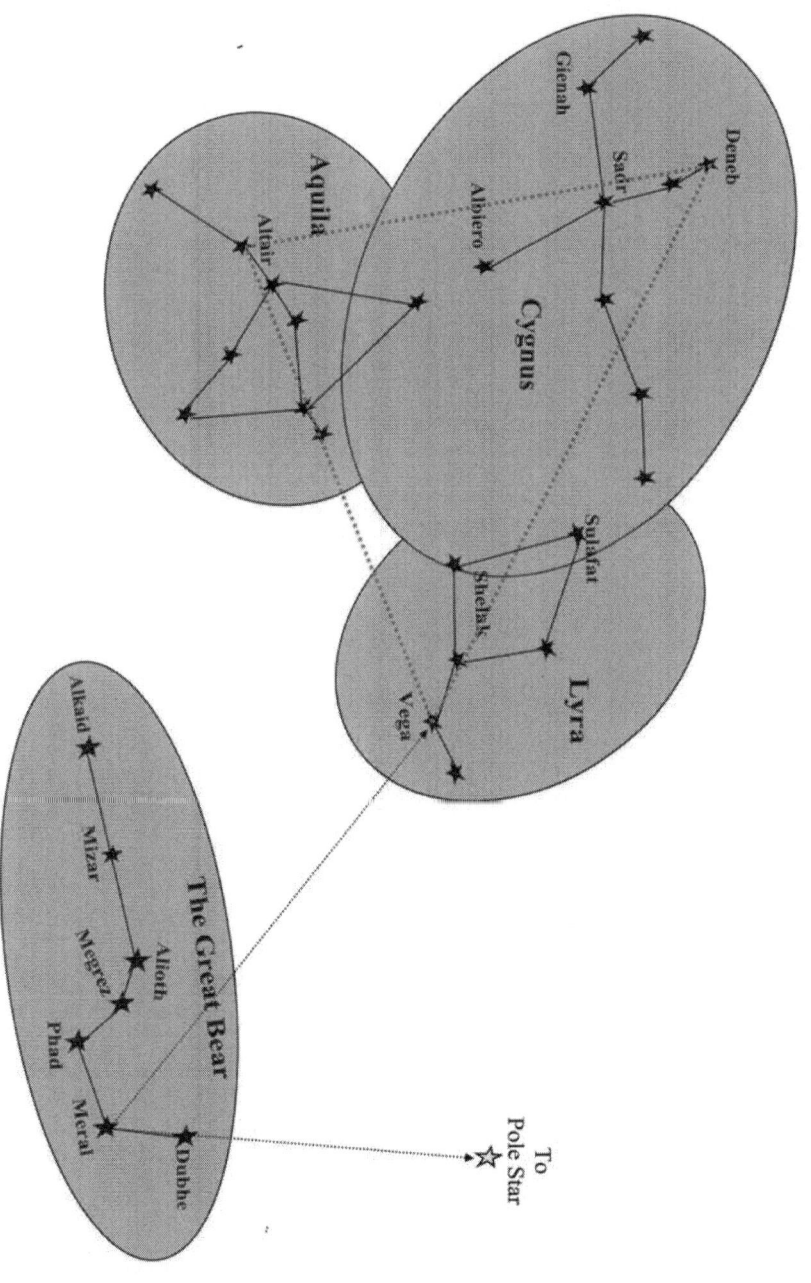

# Constellations of the Summer Triangle.
## Lyra The Harp

Lyra is a constellation in the northern hemisphere and is visible between latitudes 90°N and 40°S. From a navigator's point of view, it is best seen during nautical twilight in the late summer and autumn.

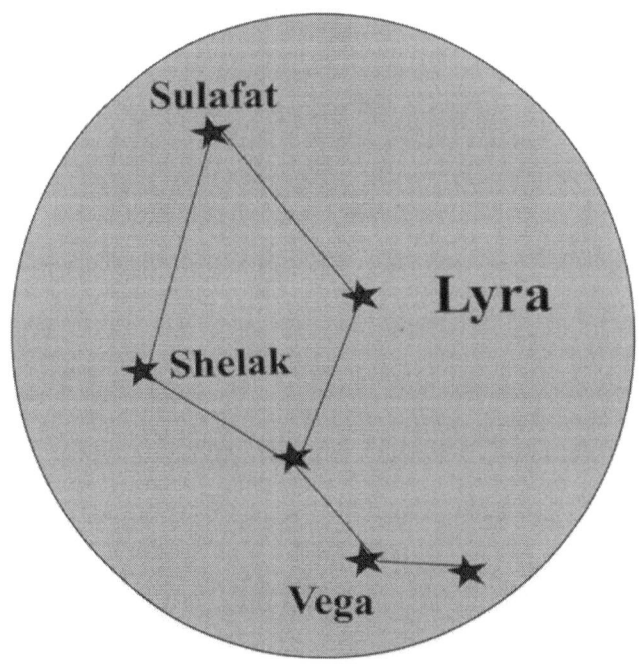

The constellation Lyra contains **Vega,** which is the second brightest star in the northern hemisphere and is a **navigational star.**

In Greek Mythology, when Orpheus died, he dropped his lyre into a river from where it was retrieved by an eagle sent by Zeus. Zeus then sent both the lyre and the eagle into the sky as the constellations Lyra and Aquila.

**Aquila, The Eagle**

Aquila is visible between latitudes 85°N and 75°S and is also best seen during nautical twilight during the late summer and autumn. The constellation Aquila contains **Altair,** the 12$^{th}$ brightest star in the sky and also a **navigational star.**

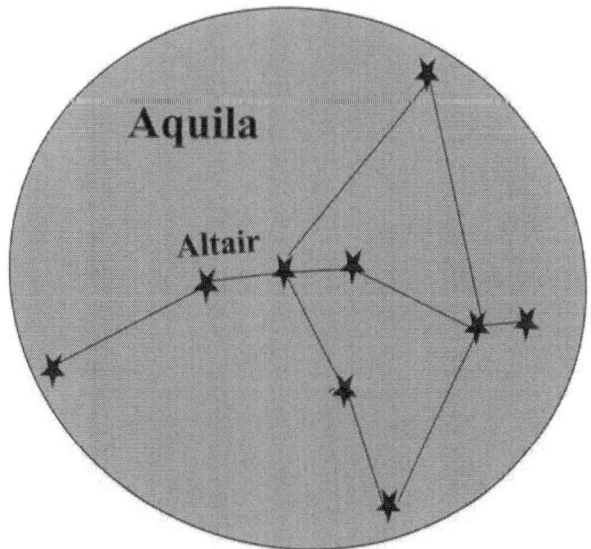

In Greek mythology, Aquila, the eagle, carried thunderbolts for Zeus and as explained above, later rescued the lyre of Orpheus from the river.

## Cygnus The Swan (The Northern Cross)

The constellation Cygnus contains 6 stars, the brightest of which is **Deneb**, the 19th brightest star in the sky and **a navigational star.** As with Lyra, Cygnus is visible between latitudes 90°N and 40°S. For navigators it is best seen during nautical twilight in late summer and autumn. Cygnus contains an asterism formed by the brightest stars in the constellation which is named the Northern Cross.

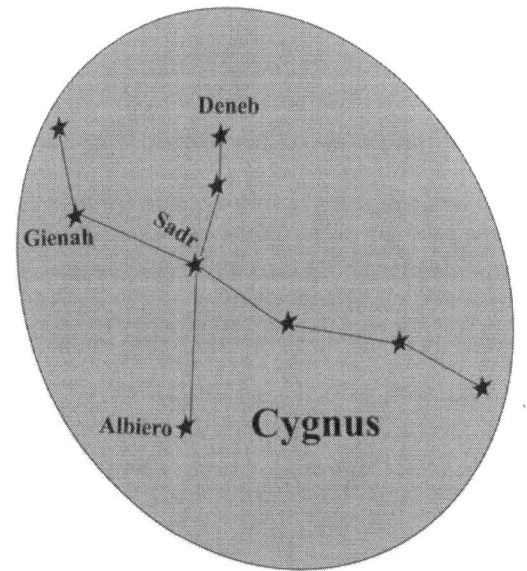

In Greek mythology, Orpheus was said to have been turned into a swan by Zeus and sent into the sky as the constellation Cygnus along with Aquila and Lyra.

## PISCES, The Fish

Pisces is a large 'V' shaped constellation which straddles the equator and lies on the path of the ecliptic. It is visible between latitudes 90°N and 65°S; it is best seen in November. The brightest star in Pisces is Alpherg or Kullat Nunu but this is not a navigational star; in fact, this constellation contains no navigational stars.

The reason that we have included Pisces in this 'route map' is because of its association with the 'First Point of Aries' which is the point at which the Sun crosses the celestial equator when it is moving from south to north along the ecliptic. This event occurs on 21/22 March and is known as the vernal Equinox. The confusing thing is that, although the 'First Point of Aries' lay in the constellation of Aries when it was chosen by the ancient astronomers, due to precession, it now lies in Pisces.

This diagram shows the arrangement of the stars contained in the constellation Pisces

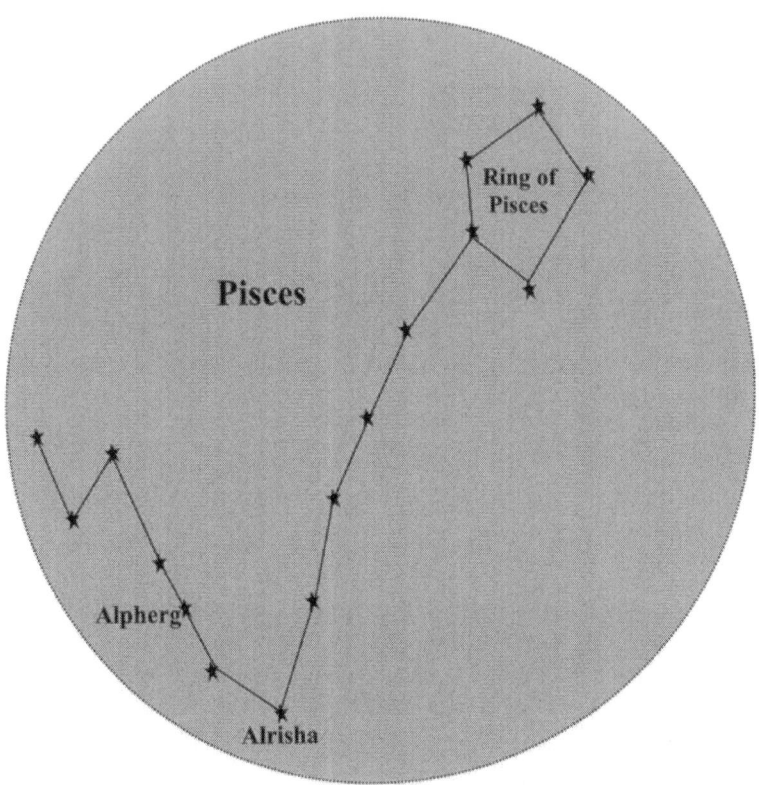

The name Pisces is derived from the Latin for fish and is said to depict two fish, swimming in opposite directions, held together by a piece of string connecting their tails. The star Alrisha is said to be the knot that ties the strings that hold the two fish together. In ancient Greek mythology, Pisces is associated with the fish that carried Aphrodite and Eros to safety from the monster Typhon. In another mythological tale, the fish of Pisces were said to have been spawned by the 'Great Fish' in the constellation Pisces Austrinus which is known as the 'Southern Fish'.

### Finding Pisces.
Pisces lies just to the south of the 'Great Square of Pegasus as shown in the diagram which follows. If a line is drawn from Scheat to Algenib in Pegasus

and extended by about one hand-span, it will point to the star Alrisha in Pisces; however, this constellation is very hard to find because it is so faint.

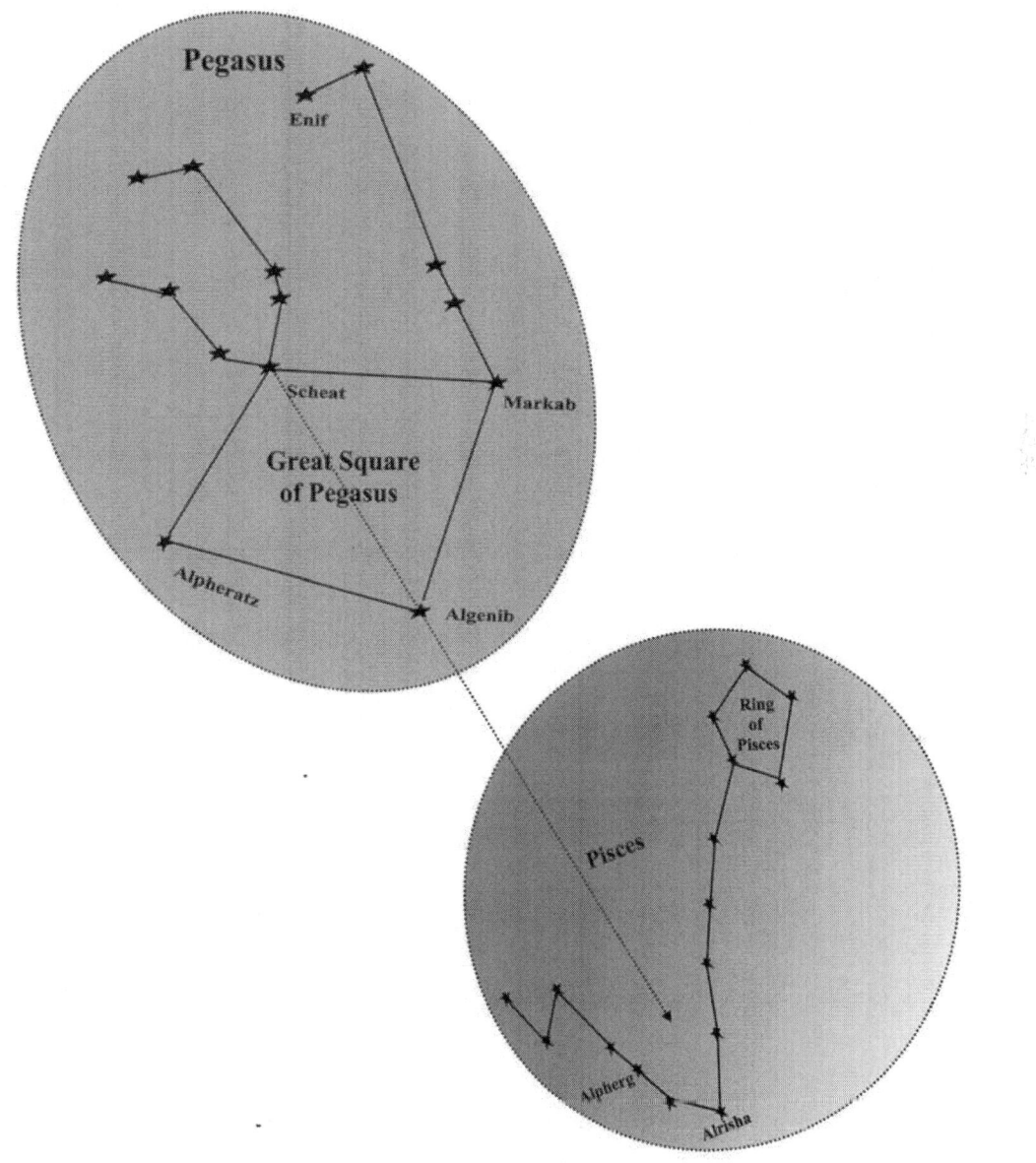

# Aquarius, the Water-Bearer

Aquarius is a constellation in the southern hemisphere and is visible at latitudes between 65°N and 90°S; it is best seen during the month of October. The brightest star in Aquarius is Sadalsuud an Arabic phrase meaning "luck of lucks". Sadalsuud is not a navigational star and in fact, there are no navigational stars in the constellation Aquarius which, like Pisces, is very faint and difficult to see with the naked eye.

In ancient Greek mythology, Zeus transformed himself into an eagle (Aquila) to carry a young man named Ganymede to serve as a cup-bearer to the gods in Olympus. The name Aquarius is derived from the Latin for 'water-bearer' or 'cup-bearer'.

## Finding Aquarius.

As stated above, Aquarius is a very faint constellation and is difficult to locate. However, this diagram shows that if we line up the two stars that form the base of the triangle at the top of the 'Ring of Pisces' and extend that line it will point to the star Sadalmeilk in the constellation Aquarius which is to the south of Pegasus.

If we also run a line from Scheat to Markab in Pegasus and extend that line by a palm-width, that too will point to Aquarius.

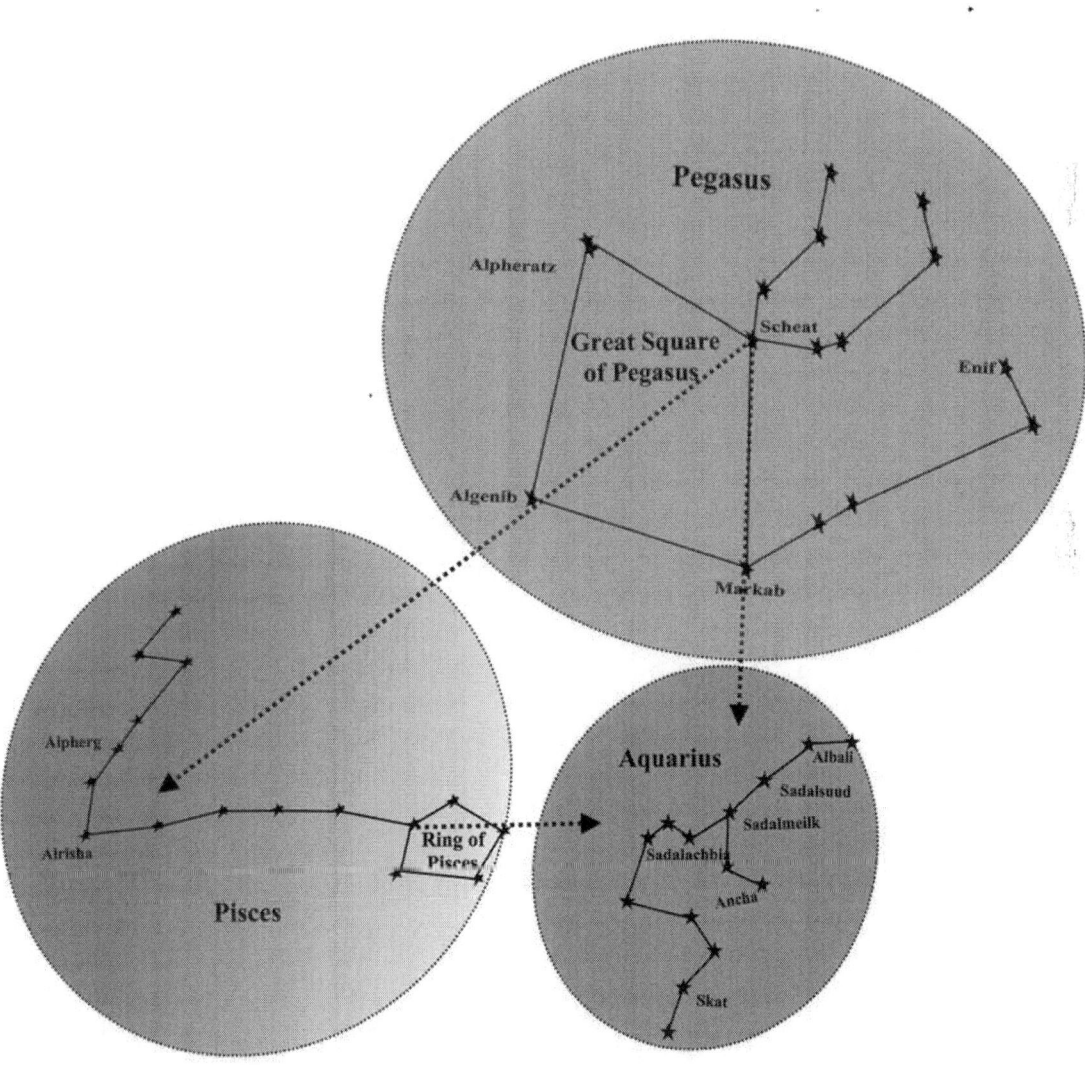

In a popular song, the words the 'dawning of the age of Aquarius' refer to the period when the vernal equinox will lie inside the constellation Aquarius. As discussed earlier, the vernal equinox is the point where the Sun crosses the Equator on its northward movement along the ecliptic and heralds the first day of spring in the northern hemisphere on 20th/21st March. This point is known as the 'First Point of Aries' because in 150 B.C. when Ptolemy first mapped the constellations, Aries lay in that position. However, although still named the 'first point of Aries', due to precession, the vernal equinox now lies in the constellation Pisces, so logically, it should be named the 'first point of Pisces' since we are now in the 'Age of Pisces'. There are various predictions of when the next 'age of Aquarius' will begin but the most prominent of these is about 2600 A.D.

## Piscis Austrinus, the Great Fish or the Southern Fish

The small constellation Piscis Austrinus, also called Piscis Australis, lies in the southern hemisphere and is visible between latitudes 55°N and 90°S. It contains mostly faint stars except for **Formulhaut** which is one of the brightest stars in the sky and is a navigational star. For navigators, the best time to see Formulhaut during nautical twilight is in the month of October.

Piscis Austrinus is associated with the Babylonian myth about the goddess Atargatis who fell into a lake and was rescued by a large fish. The ancient Greeks named the constellation the 'Great Fish' which, according to Egyptian mythology, saved the goddess Isis who, as a reward, sent it into the sky where it spawned the two fish in the constellation Pisces. The name Piscis Austrinus is derived from the Latin for the 'Southern Fish'.

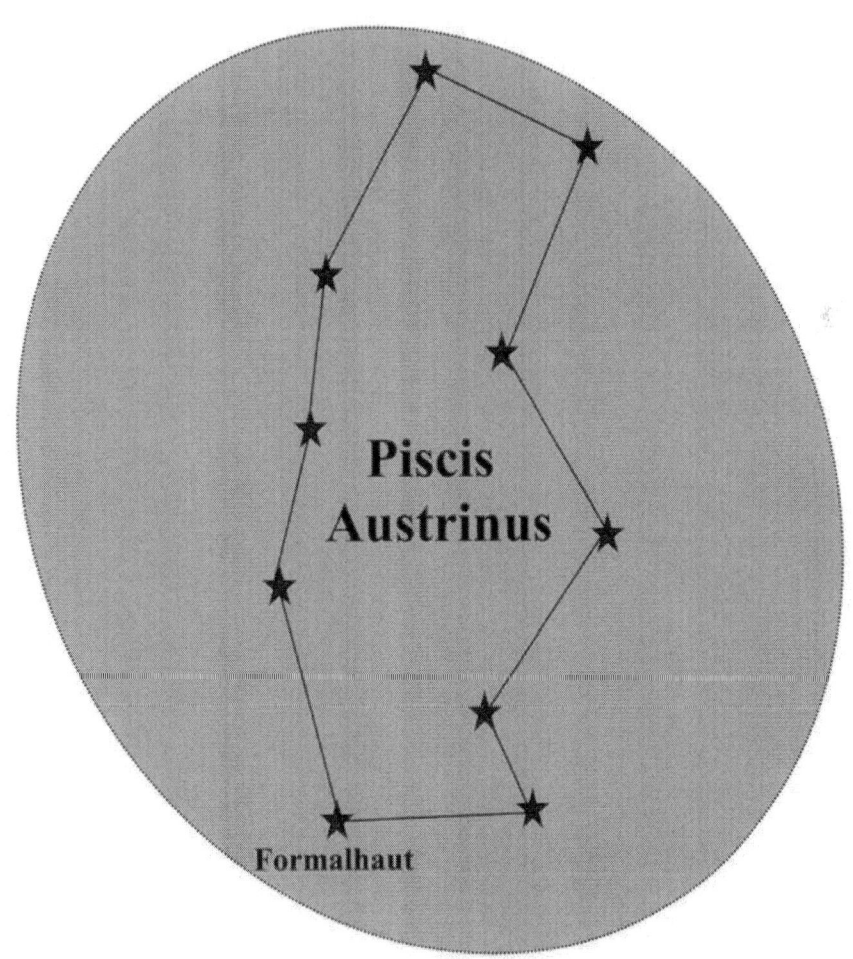

# Finding Piscis Austrinus

Fomulhaut was once considered to be part of Aquarius as well as the constellation Pisces Austrinus, where it now belongs. Formulhaut is depicted as the toe of Aquarius and this idea provides us with a way of locating both Aquarius and Pisces Austrinus for if we can find Formulhaust, the brightest star in the region, then we can find both of those constellations.

The following diagram shows Pisces Austrinus nestling at the foot of Aquarius with Formulhaut providing the link between them.

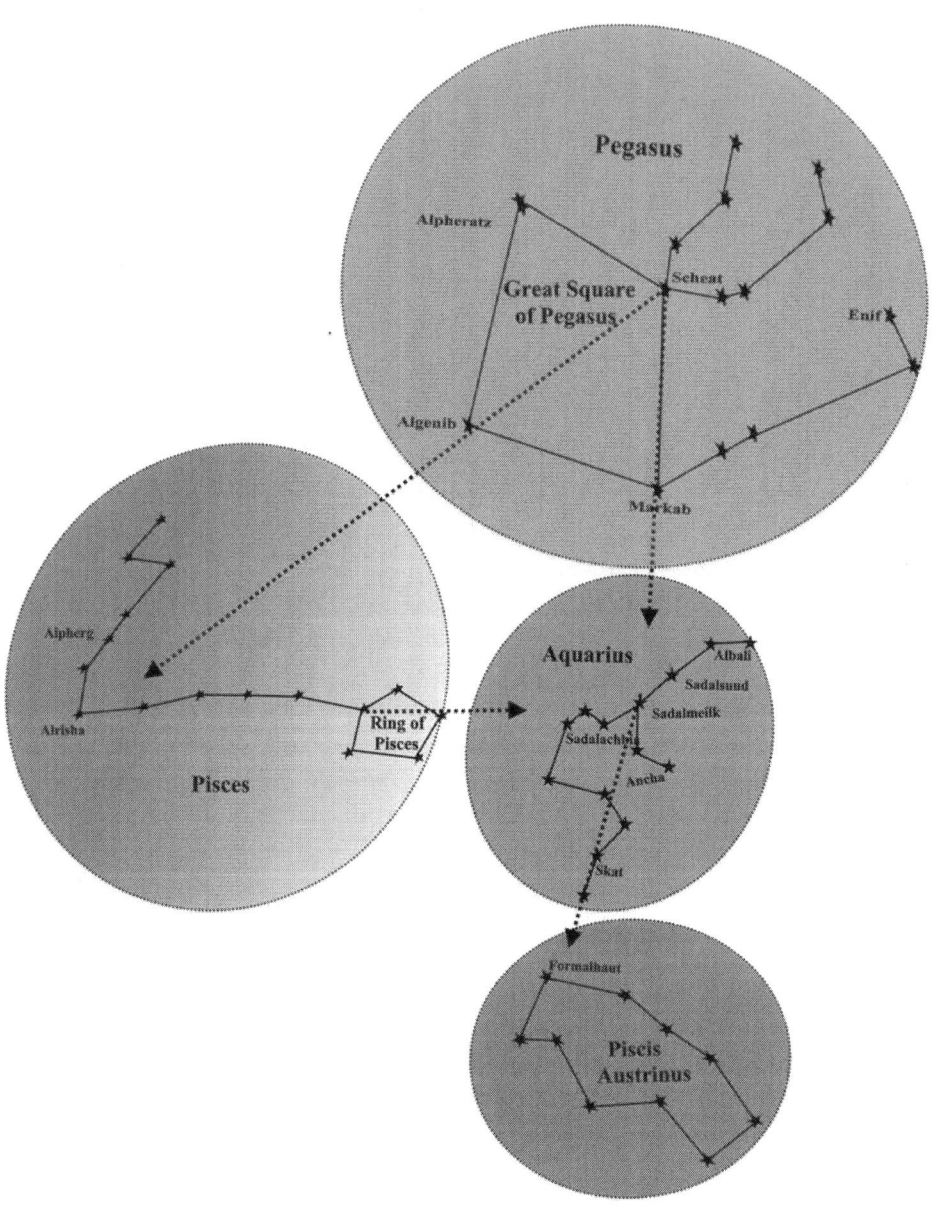

# Capricorn (Capricornus) The Goat

The constellation Capricornus is in the southern hemisphere and is visible between latitudes 60°N and 90°S; it is best seen in the early evening in September.

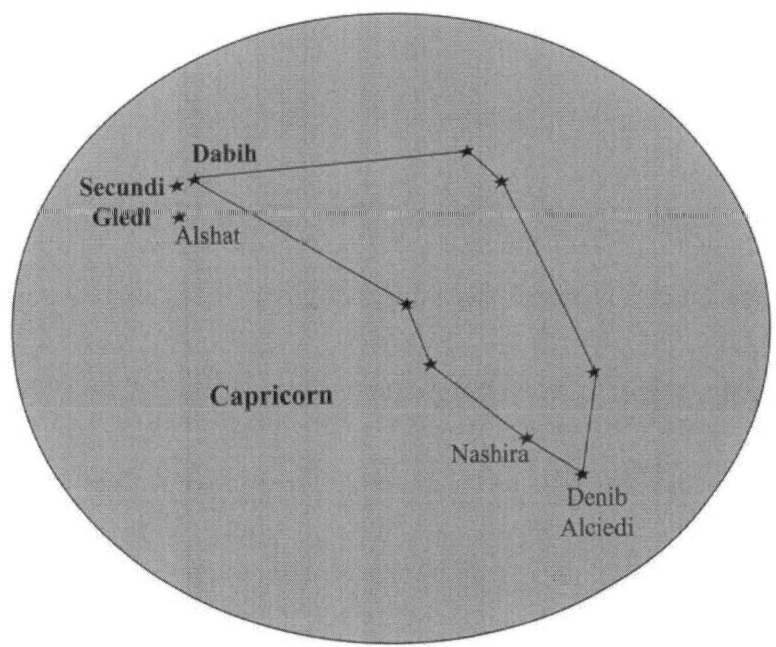

The tropic of Capricorn is so named because in ancient times, the Sun passed through the constellation Capricornus during the Winter Solstice on 21/22 December, when the Sun's declination reached its southernmost latitude of 23.4°S. However, due to precession, the Sun is now over the constellation Sagittarius at the Winter Solstice.

Although Capricornus has several bright stars that form the shape outlined in the image above, it is one of the faintest constellations in the sky and contains no navigational stars.

Capricornus is associated with many mythological tales. In Greek mythology, one story is that Pan transformed himself into a goat and jumped into the river Nile to escape from the monster Typhon.

## Finding Capricornus

As stated above, Capricornus is one of the faintest constellations in the sky and for this reason it is difficult to find; however, as shown in the drawing below, the Summer Triangle can help us to find it.

If we imagine a line running from Vega in the constellation Lyra to Altair in the constellation Aquilla and extend that line by about one and a half hand-spans, it will point to Capricornus.

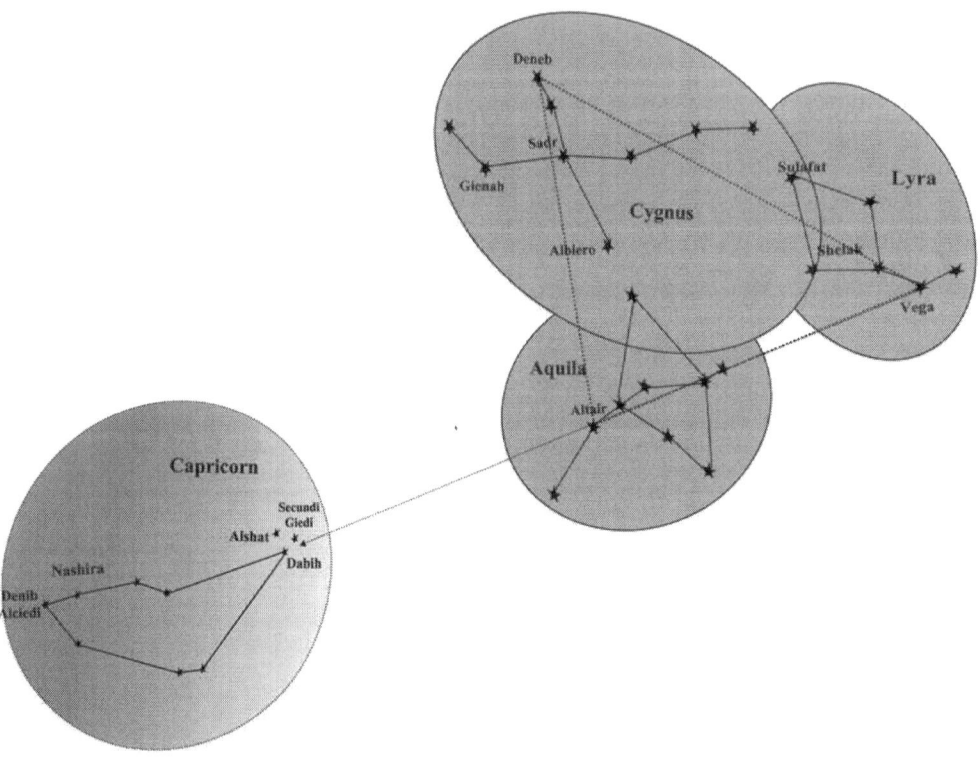

# Taurus, The Bull.

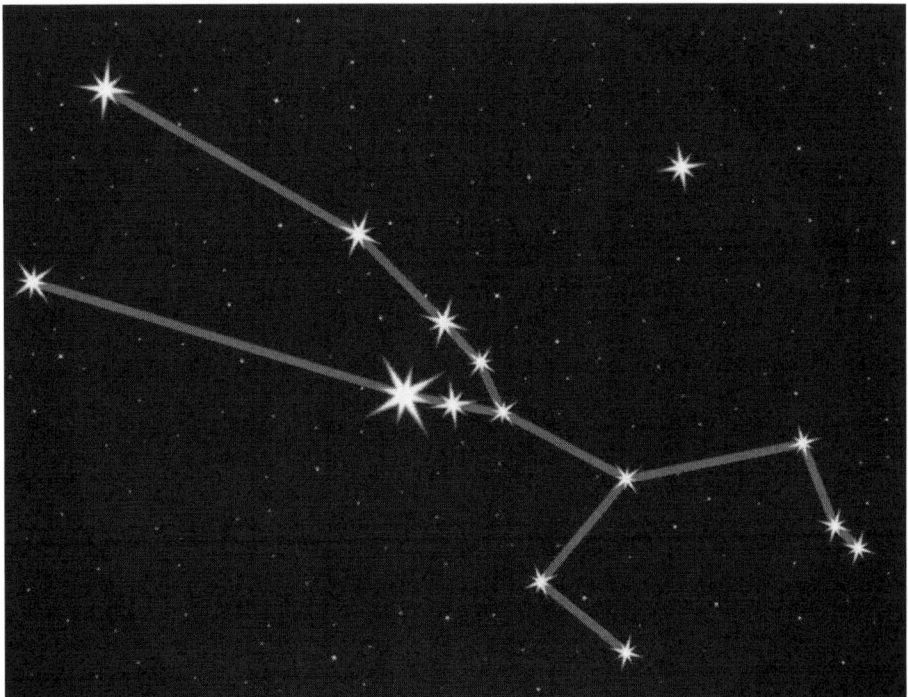

Taurus is a constellation in the northern hemisphere and is visible at latitudes from 90°N to 65°S; for navigators, it is best seen during nautical twilight in January.

**Aldebaran**, which is known as Taurus' Eye, is the 14th brightest star in the sky and is an important navigational star. The star at the tip of the northern horn of Taurus, **Al Nath** (sometimes spelled El Nath) is the second brightest star in the constellation and is also a navigational star. In earlier times this star was considered to be shared with the constellation Auriga, forming the right foot of the Charioteer as well as the Northern horn of the bull.

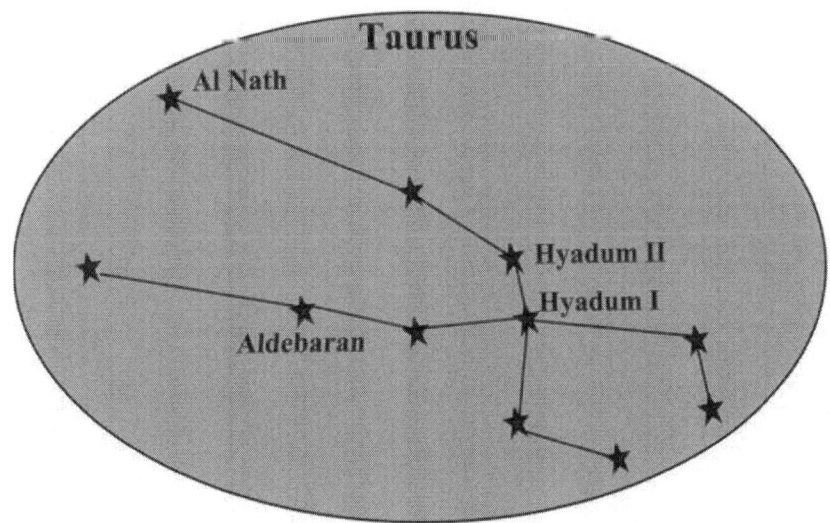

Taurus is associated with several mythological beliefs. In Greek mythology, Zeus was said to have disguised himself as a bull to abduct Europa, the daughter of king Agenor. The ancient Egyptians believed that the constellation represented the sacred bull associated with spring and Babylonian astronomers called it the 'Heavenly Bull'.

## Finding Taurus.

If we imagine a line from Phad to Meral in Ursa Major and extend it for a distance of 80° or roughly 4 hand-spans it will point directly to the star Aldebaran in the constellation Taurus. Once we have located Aldebaran the remaining stars of Taurus can easily be identified.

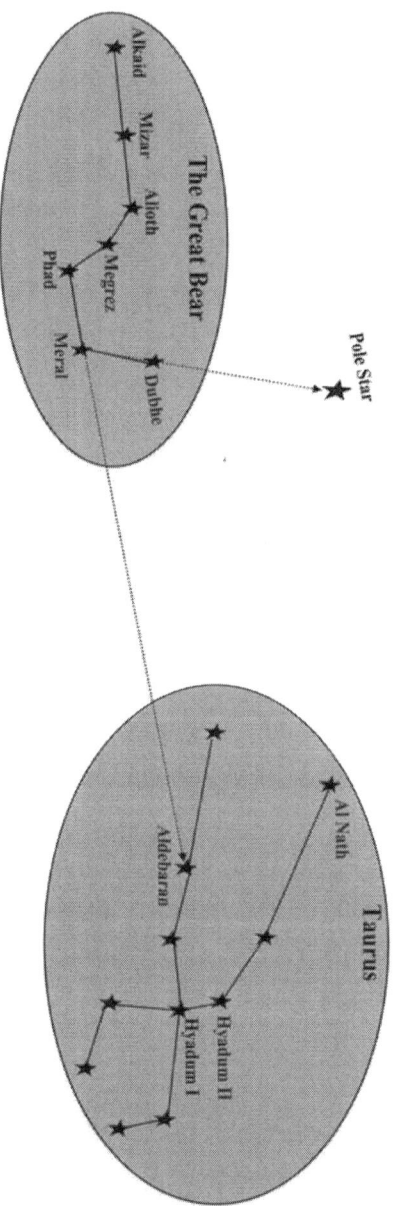

# The Pleiades, the 'Sailing Stars' or the 'Seven Sisters'.

The Pleiades form a small star cluster close to Taurus; it is visible between latitudes 90°N and 65°S and is best seen during the month of January. Merope is the brightest of these stars but even so, it not considered to be a navigational star. In ancient times, the Pleiades were called the Sailing Stars because Greek sailors would not put to sea unless they could be seen in the sky.

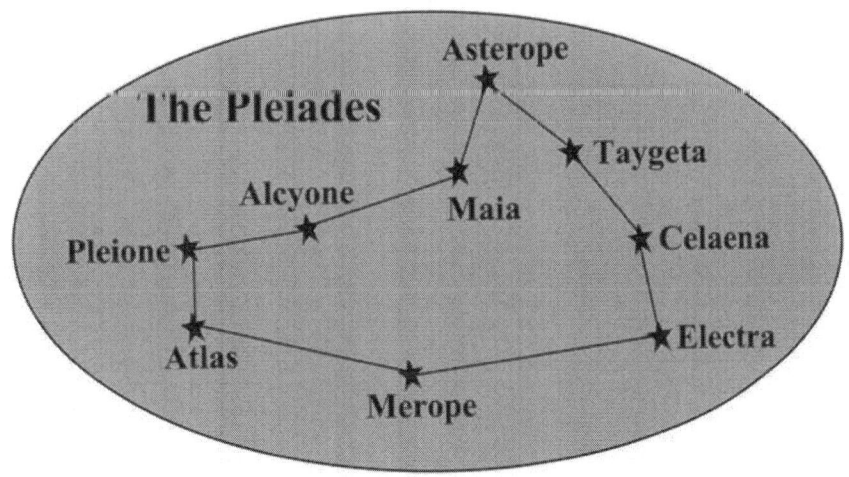

Another name for the Pleiades is the 'Seven Sisters' after Alcyone, Maia, Anterope, Taygeta, Celaena, Electra and Merope from Greek Mythology. Although Pleione was not one of the seven sisters, she along with her consort Atlas is included in the Pleiades Cluster.

## Finding The Pleiades

The next diagram shows that if, in the constellation Taurus, we draw a line from Aldebaran to Tau Tauri (the two eyes of the Bull) this will point roughly in the direction of the Merope in the star cluster Pleiades which lies about 10° (a palm-width) from Taurus.

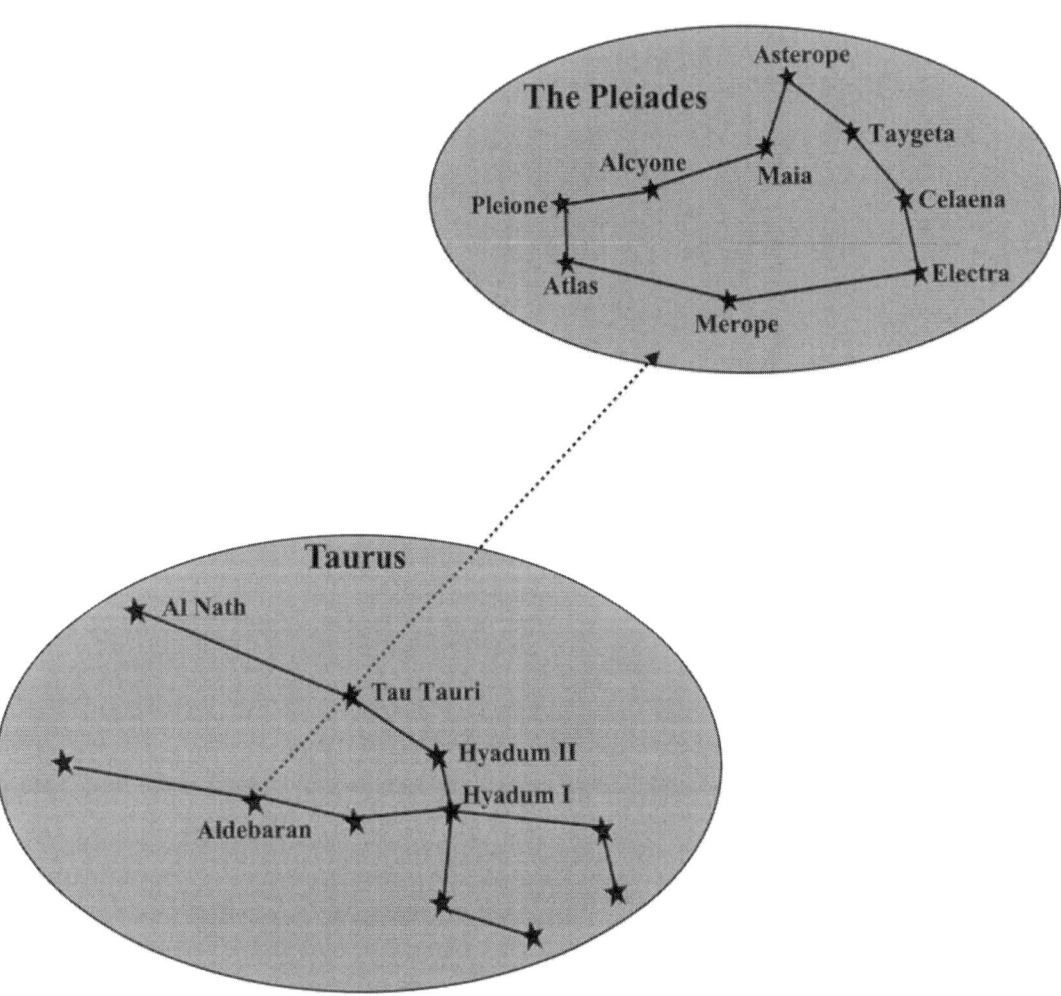

# Auriga the Charioteer

The constellation Auriga is in the northern hemisphere and is visible between latitudes 90°N to 40°S. It can be seen during nautical twilight in February and March.

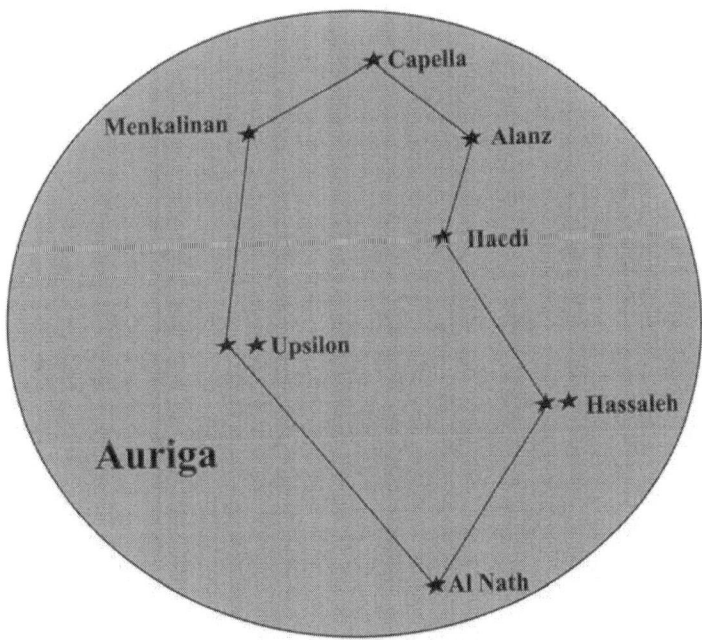

**Capella** is one of the navigational stars; it is the brightest star in the constellation Auriga and the 6th brightest in the northern hemisphere.

The diagram below shows the star Al Nath (sometimes spelled El Nath) forming the right foot of the Charioteer and as has already been explained, it was once considered to be shared with the constellation Taurus where it forms the northern horn of the bull. When the constellation boundaries were changed in 1930, Al Nath was assigned solely to Taurus. However, if we still think of Al Nath (the second brightest star in Taurus and also the second brightest in Auriga) as the foot of Auriga, we will have an easy method of finding both Auriga and Taurus as we can see from the diagram below.

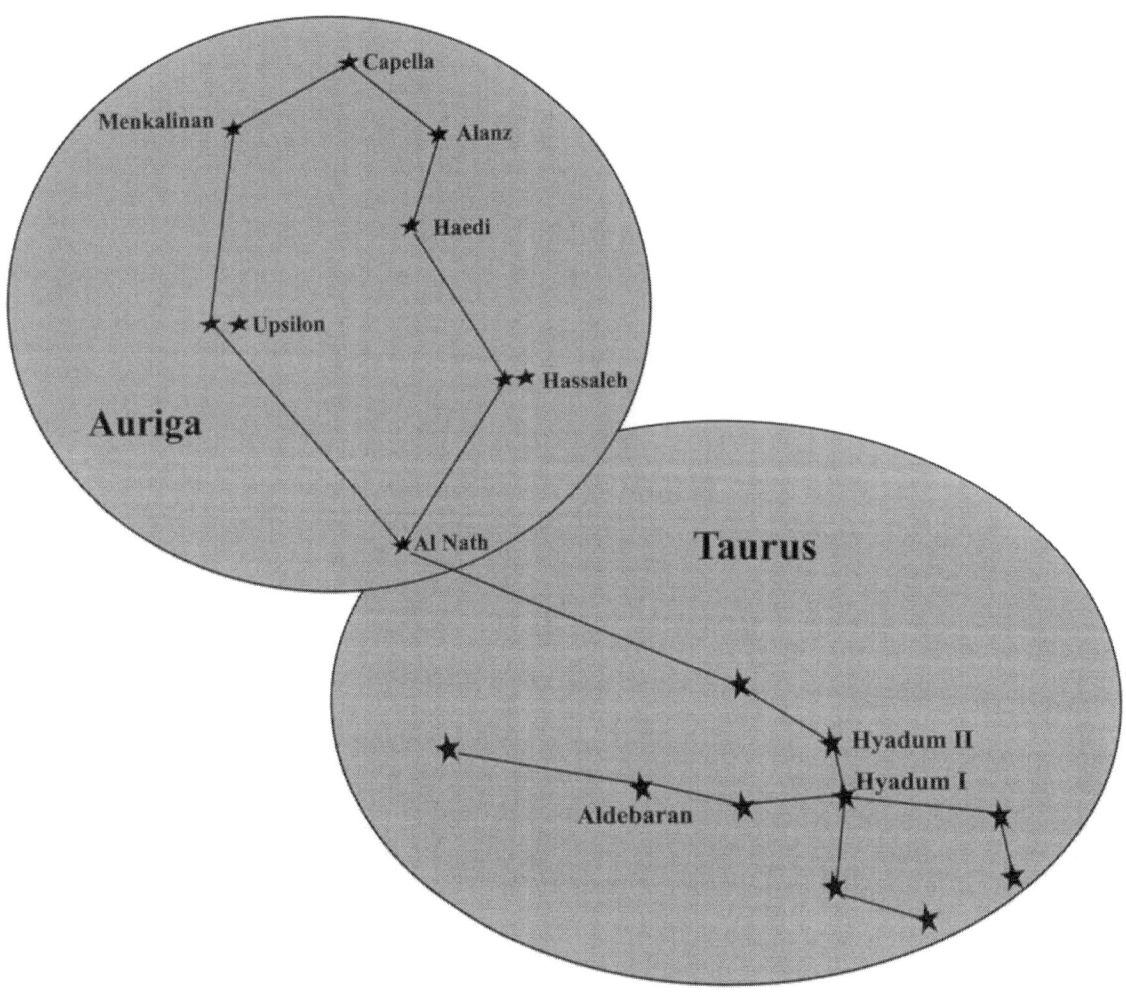

In mythology, Auriga is associated with Myrtilus the charioteer because the shape of the constellation was said to resemble a pointed helmet of a charioteer. It is also identified with Hephaestus, the god of the blacksmiths who invented the chariot.

## Orion, The Hunter

Orion which lays below Taurus straddles the celestial equator and is visible between latitudes 95°N and 75°S. This easily recognized constellation contains 4 navigational stars, **Rigel, Belatrix, Anilam** and **Betelgeuse** which, for navigation purposes are best seen during nautical twilight in the month of January.

The Orion constellation is named after The Hunter in Greek mythology. Meissa marks the position of the Hunter's head while Betelgeuse, and Belatrix are his shoulders. Alnitak, Alnilam and Mintaka form his belt and from this hangs his sword which is marked by the Orion Nebula. His right thigh is marked by Saiph and Rigel marks his left foot.

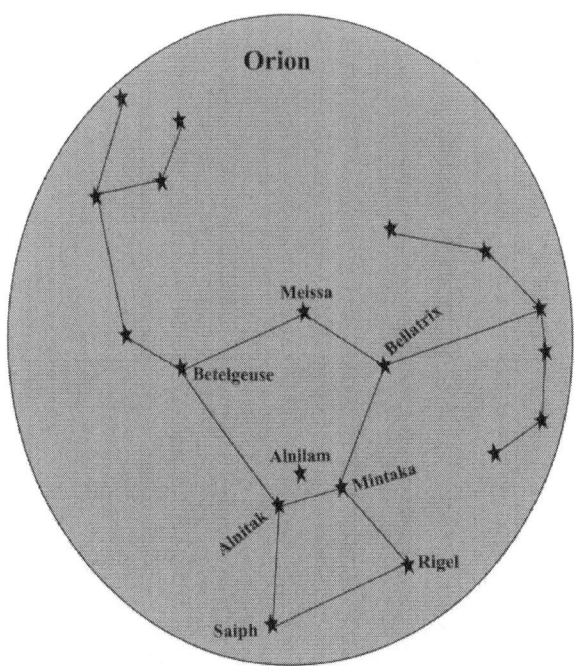

## Finding Orion

If we take a line from Tau Tauri to Aldebaran in the constellation Taurus and extend this line for roughly three hand-spans we will come to the star Belatrix in Orion as the diagram below shows.

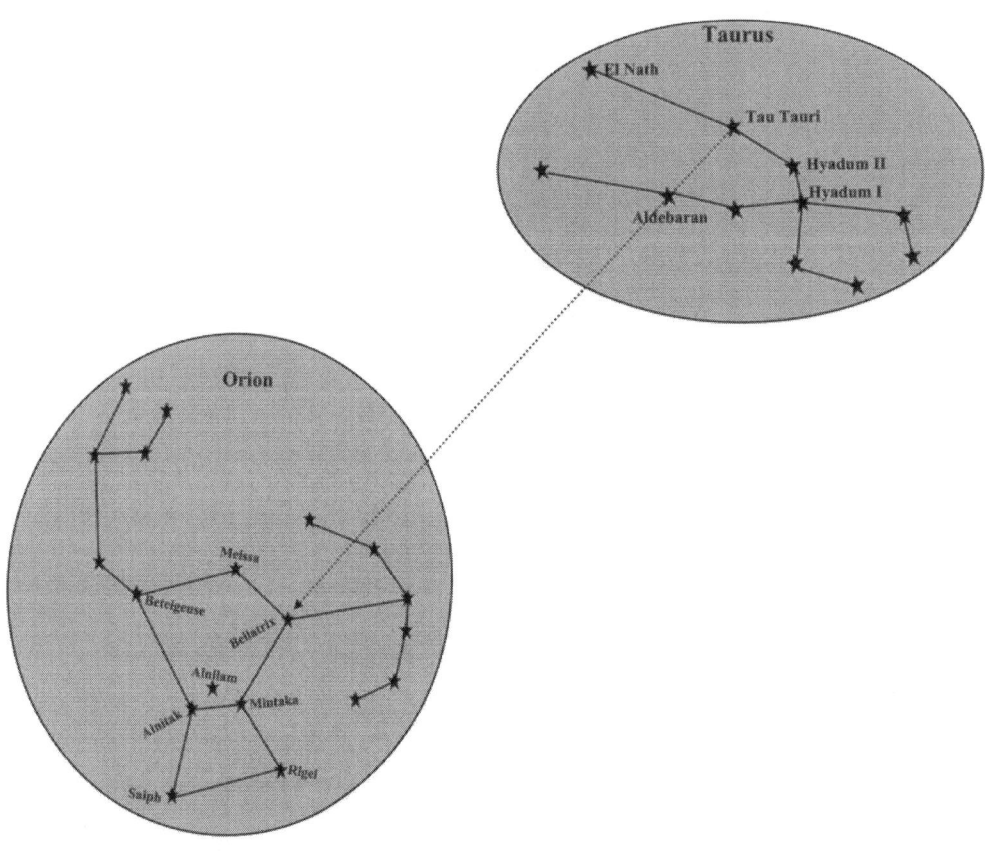

# Canis Major, The Greater Dog

The constellation Canis Major is in the Southern Hemisphere; it is visible from 90°S to 60°N and is best visible from November to March.

Canis Major is Latin for 'The Greater Dog' and is so named because it was said to represent one of the two hunting dogs of Orion the Hunter; the other dog being Canis Minor.

In Greek mythology, Zeus sent Laelaps, an alternative name for Canis Major, into the sky when it failed to outrun a fox.

As shown in the diagram below, Canis Major contains Sirius, which is otherwise known as the Dog Star. **Sirius** is the brightest star in the sky and is a navigational star which for navigation purposes is best seen during nautical twilight during the month of February.

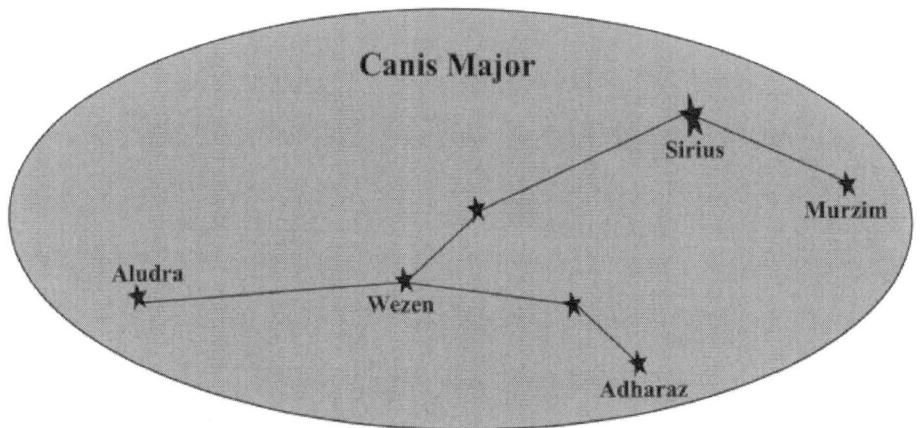

## Canis Minor The Lesser Dog

Canis Minor is a small constellation in the northern hemisphere and is visible at latitudes from 85°N to 75°S. It is best seen during nautical twilight in March.

Its name means 'the lesser dog' in Latin and it is said to represent one of the dogs that follow Orion. In another mythological tale, Canis Minor represents Maera the dog which was sent to the sky by Zeus after it died of grief when its master Icarius was killed.

As shown in the next diagram, there are only two bright stars in Canis Minor, Gomeisa and **Procyon** which is a navigational star.

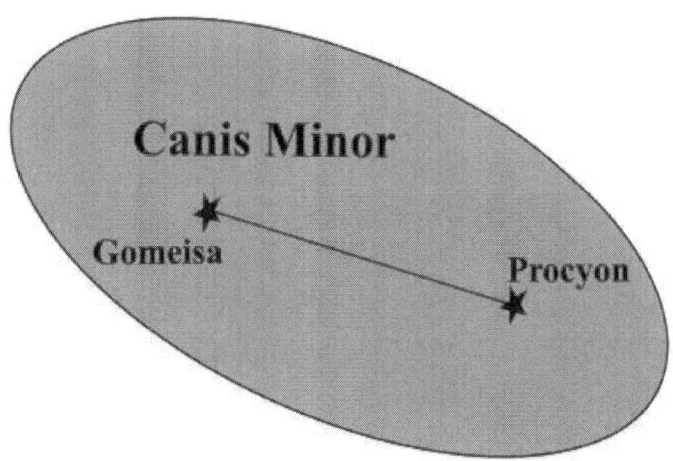

### Finding Canis Major and Canis Minor

The next diagram shows the positions of the constellations Canis Major and Canis Minor in relation to Orion.

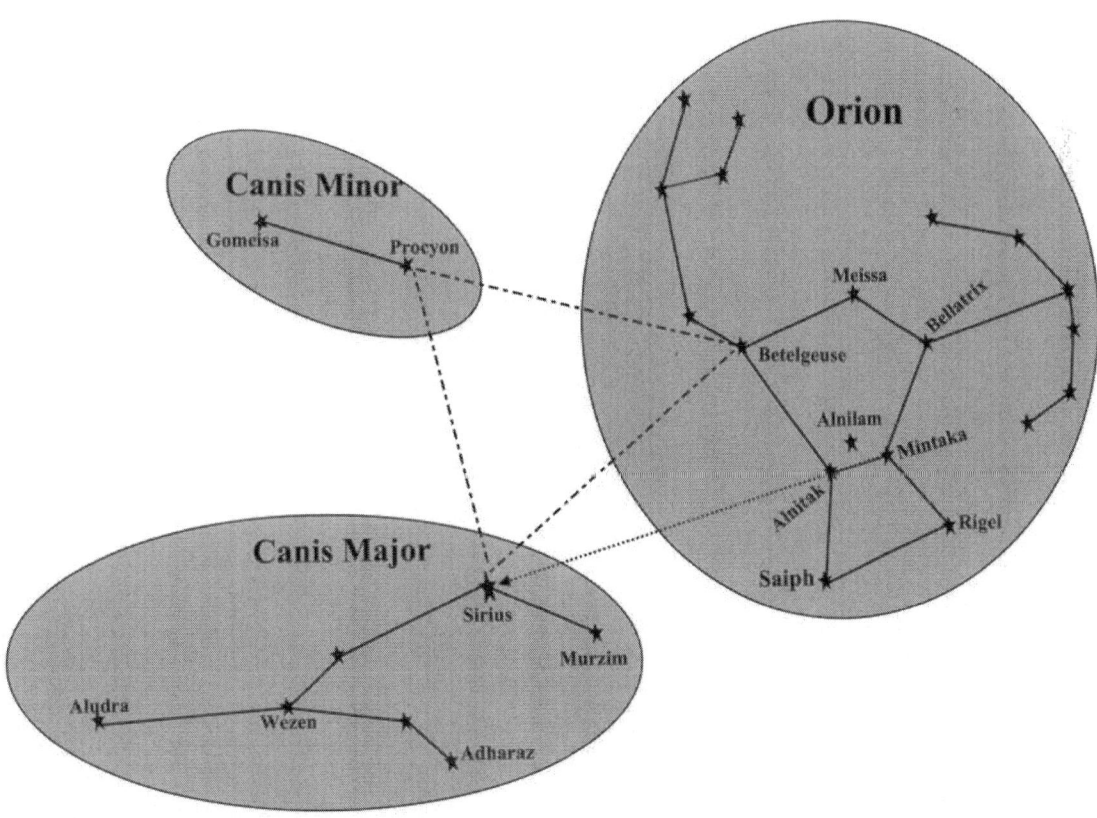

If we imagine a line from Mintaka to Alnitak in Orion and then extend that line by roughly one and a half hand-spans, it will point to Sirius the brightest star in the constellation Canis Major.

Canis Minor is slightly more difficult to find because there are no direct pointers to show its position. However, as shown in the diagram, if we take a line from Betelgeuse in Orion to Sirius in Canis Major and imagine that it forms the base of an equilateral triangle, known as the **'Winter Triangle'** or **'Southern Triangle'**, we will see that Procyon in Canis Minor sits at the apex of that triangle.

**Aries, The Ram**

The constellation Aries is located in the Northern Hemisphere and is visible between latitudes 90°N and 60°S. The name Aries, which means Ram in Latin, is associated with Jason and the Golden Fleece in Greek mythology.

Aries is a difficult constellation to see with the naked eye but it can be found about mid way between Pegasus and the Pleiades. Pegasus lies to its west, Pleiades to its east as the following diagram shows. (Remember that, in star maps, east and west are reversed).

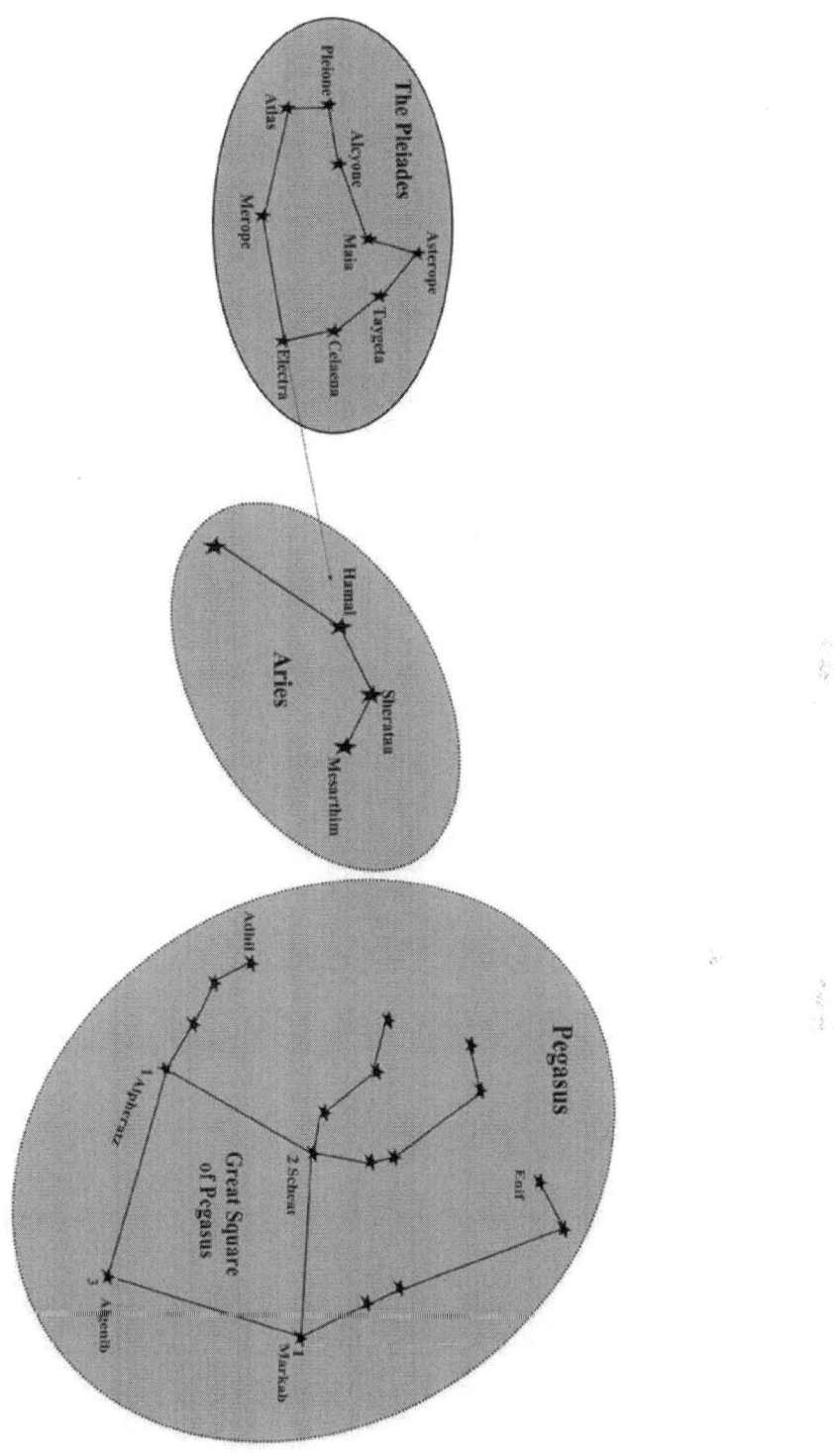

**The First Point of Aries.** Just as the Greenwich meridian has been arbitrarily chosen as the zero point for measuring longitude on the surface of the Earth, 'the first point of Aries' has been chosen as the zero point in the celestial sphere. It is the point at which the Sun crosses the celestial equator moving from south to north (at the vernal Equinox in other words). The confusing thing is that, although this point lay in the constellation of Aries when it was chosen by the ancient astronomers, due to precession, it now lies in Pisces. The First Point of Aries is usually represented by the 'ram's horn' symbol shown below:

**Hamal** is the brightest star in Aries and it is a navigational star which, for navigation purposes, is best seen during nautical twilight in the month of December.

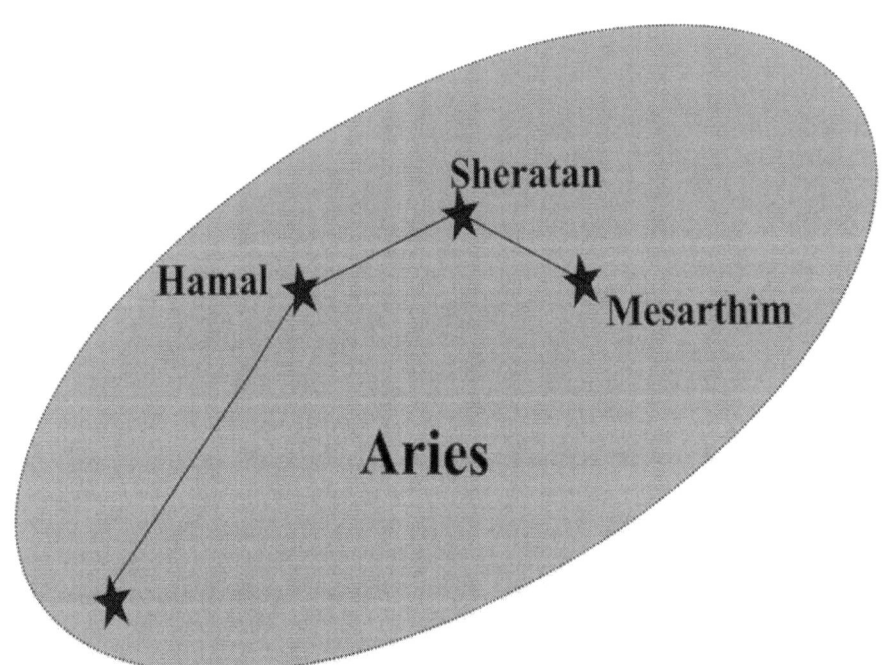

# Sagittarius, The Archer

Sagittarius is a large constellation lying over the southern hemisphere and is visible between latitudes 55°N. and 90°S. It contains several bright stars including two navigational stars, **Nunki** and **Kaus Australis** which are best seen during nautical twilight in August.

In ancient Greek mythology, Sagittarius was said to represent the Archer, a beast called a Centaur which was half man and half horse. In this representation, the Archer has a drawn bow with the arrow pointing to the star Antares, the heart of the scorpion, which had been sent to kill Orion.

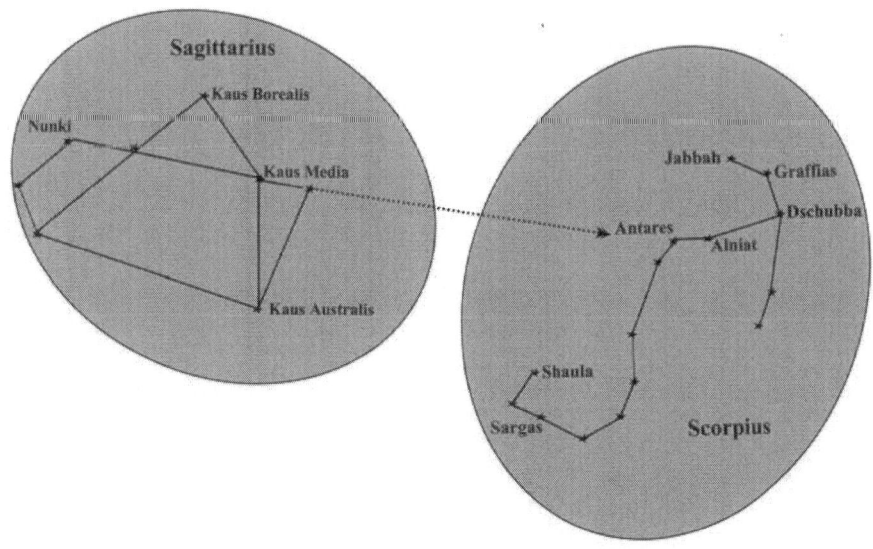

**Finding Sagittarius.** The Summer Triangle provides a useful pointer to Sagittarius. If we draw an imaginary line from the star Deneb through the star Altair in the Summer Triangle and extend that line by about 20° or one hand-span, it will point to the constellation Sagittarius as the diagram below shows.

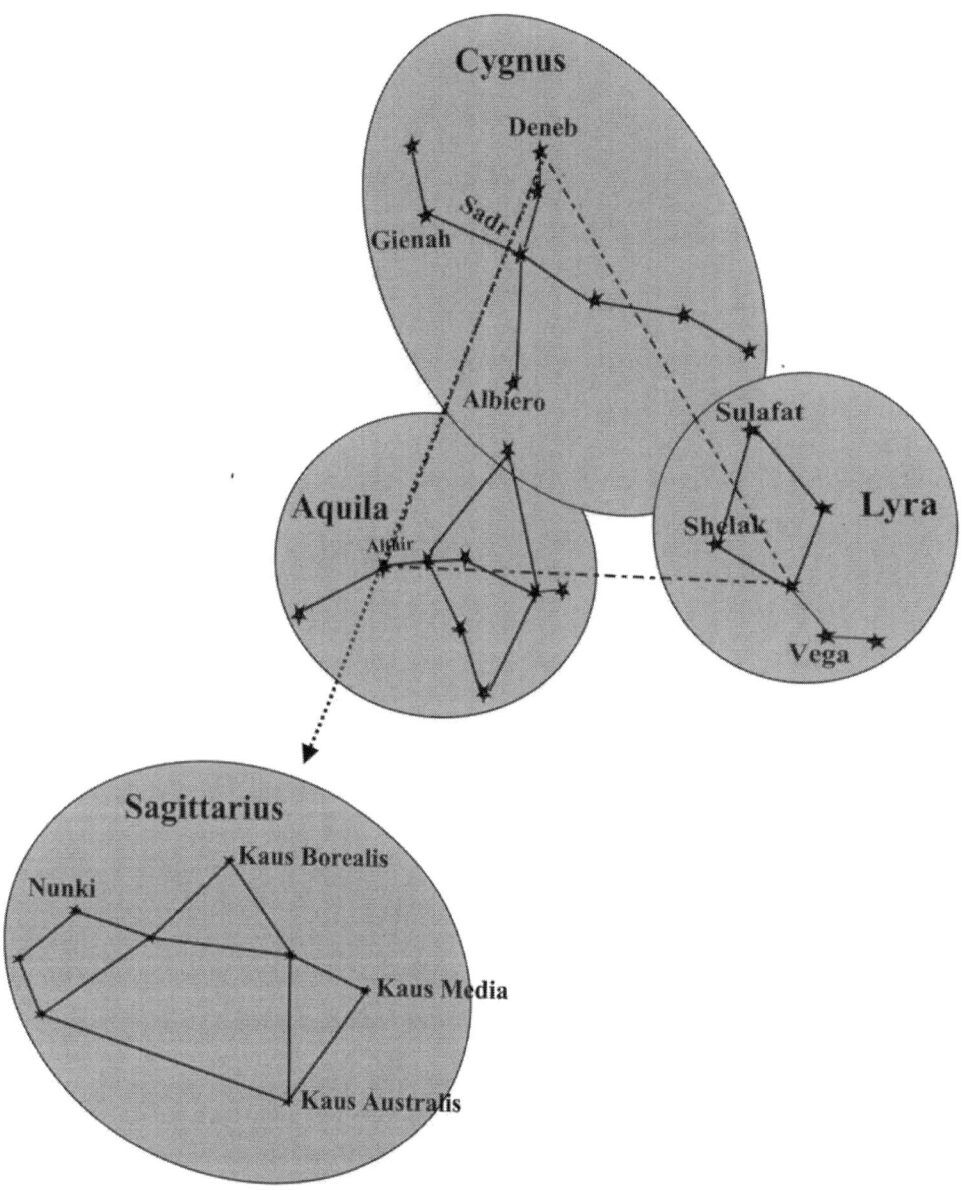

**Note.** Nowadays, the Sun is over the constellation Sagittarius at the Winter Solstice on 21/22 December when the Sun's declination reaches its southernmost latitude of 23.4° south. However, in ancient Greek times, the

Sun passed through the constellation Capricornus at this time hence the reason for naming the latitude 23.4° south the **Tropic of Capricorn**.

## Scorpius (Scorpio), The Scorpion

The constellation Scorpius lies above the southern hemisphere and is visible between latitudes 40°N and 90°S.

Scorpius has several bright stars which, between them, form the shape of a scorpion. The brightest star in Scorpius is Antares which is often mistaken for Mars because of its reddish orange colour. **Antares** is the 16th brightest star in the sky and is a navigational star. The second brightest star in Scorpius is Shaula which is said to represent the sting in the tail of the scorpion. **Shaula** is also navigational star. For navigation purposes, Antares and Shaula are best seen during nautical twilight in July.

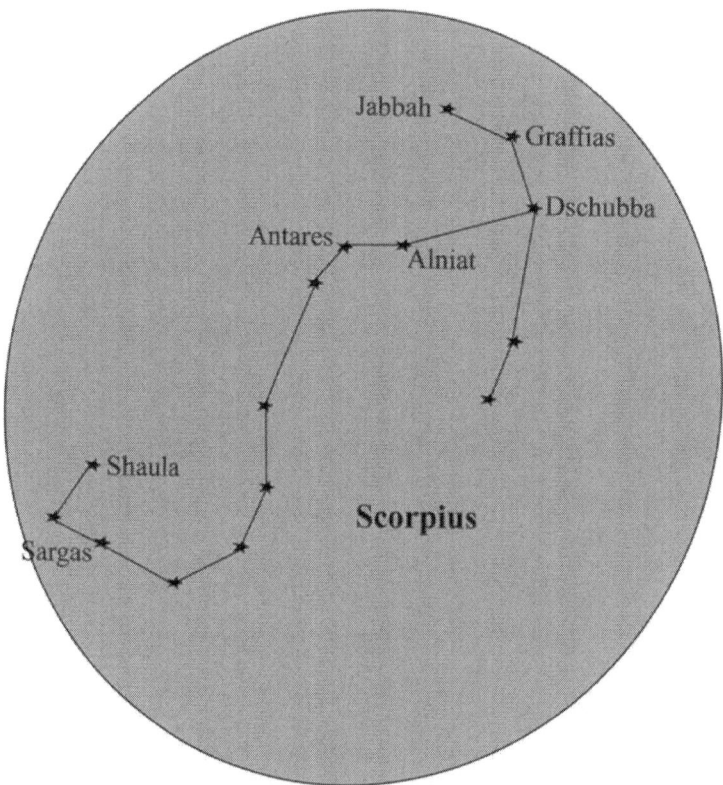

In Greek mythology, Scorpius represents the scorpion that the goddess Artemis sent to sting and kill Orion who had tried to ravish her.

## Finding Scorpius.

The legend that the Archer's arrow in Sagittarius points to Antares helps us to find and identify Scorpius. The line from Nunki to Kaus Media represents the arrow, the head of which points to Antares in Scorpio. It also helps to remember that the orange star Kaus Media points along the line of the arrow towards the red star Antares. The bright red glow of Antares further helps us to identify Scorpius.

The angular distance from Kaus Australis in Sagittarius to Shaula in Scorpius is approximately 10° or roughly equivalent to the width of the palm of the hand when held at arms length.

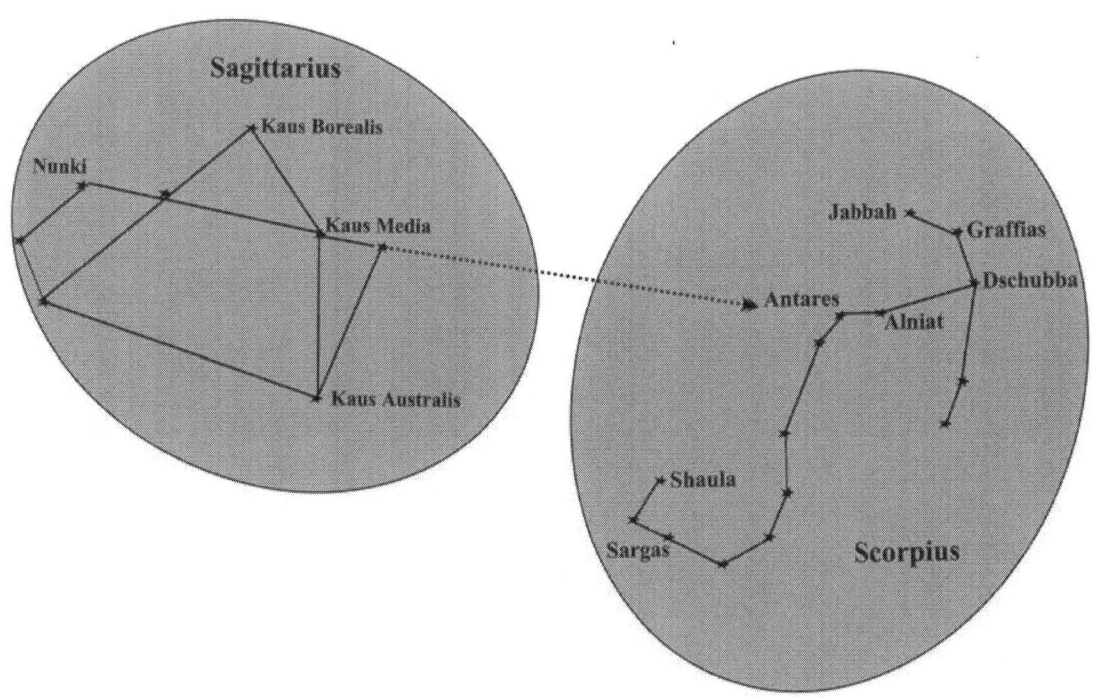

## Boötes and Corona Borealis

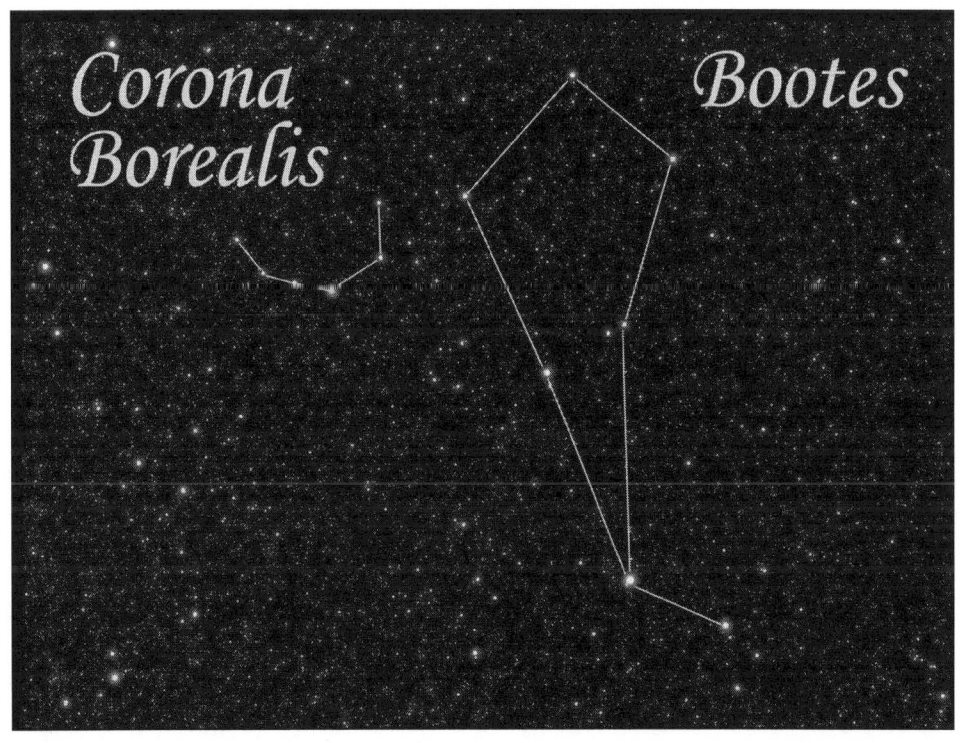

# Boötes  The Herdsman

If we take a line from Alioth to Alkaid in the Great Bear and extend that line in an imaginary curve for about roughly three hand-spans as shown in the diagram below, it will point to Arcturus, the brightest star in the constellation Boötes.

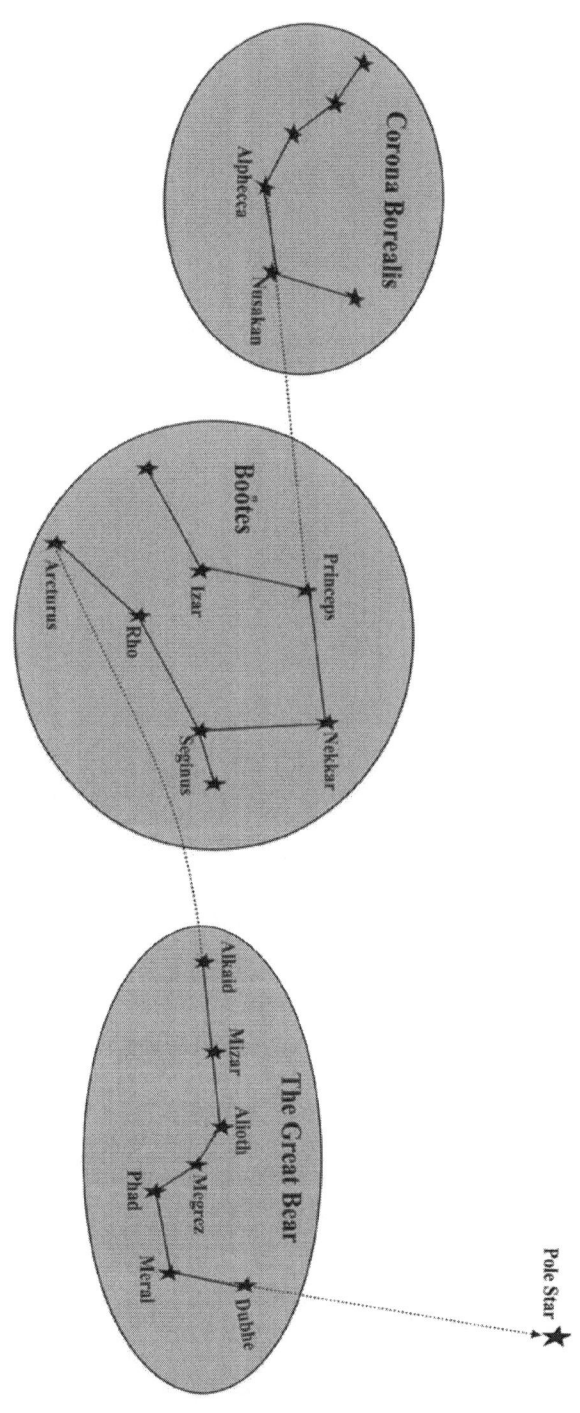

Boötes, the 13th largest constellation is located in the northern hemisphere and can be seen from 90°N to 50°S. The ancient Greeks visualised it as a herdsman chasing Ursa Major round the North Pole and its name is derived from the Greek for "Herdsman".

**Arcturus** is the fourth brightest star in the sky and is a navigational star. The ancient Greeks named Arcturus the "Bear Watcher" because it seems to be looking at the Great Bear (Ursa Major).

## Corona Borealis The Northern Crown

If we next take a line from Nekkar to Princepes in Boötes, and extend that line by about one and a half hand-spans, it will point to Nusakan in the close by Corona Borealis constellation.

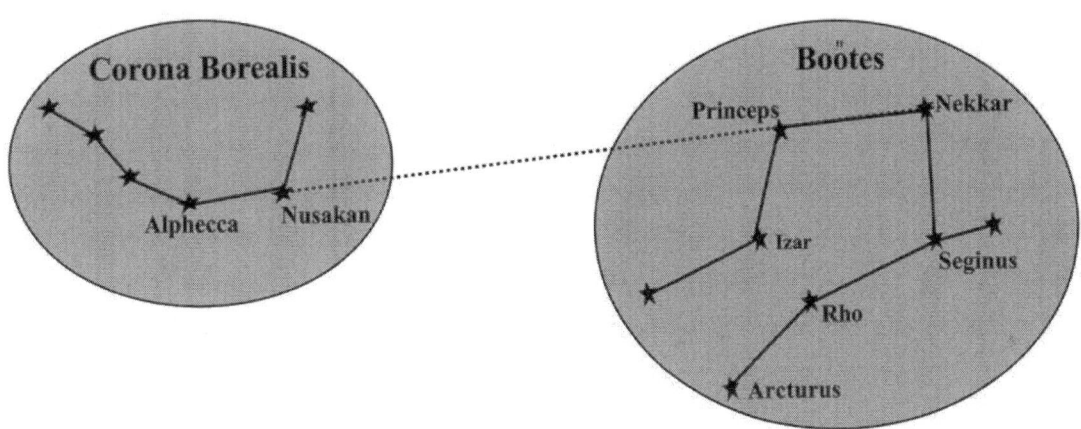

Corona Borealis, whose name in Latin means 'northern crown', is a small constellation in the northern hemisphere and can be seen between latitudes 90°N and 50°S.

The main stars in Corona Borealis form a semi-circle which is associated with the crown of Ariadne in Greek mythology. It is said that the crown was given by Dionysus to Ariadne on their wedding day and after the wedding, he threw it into the sky where the jewels became stars which were formed into a constellation in the shape of a crown.

**Alphecca**, the brightest star in the group, is a navigational star and is best seen during nautical twilight in July.

# Virgo, the Virgin

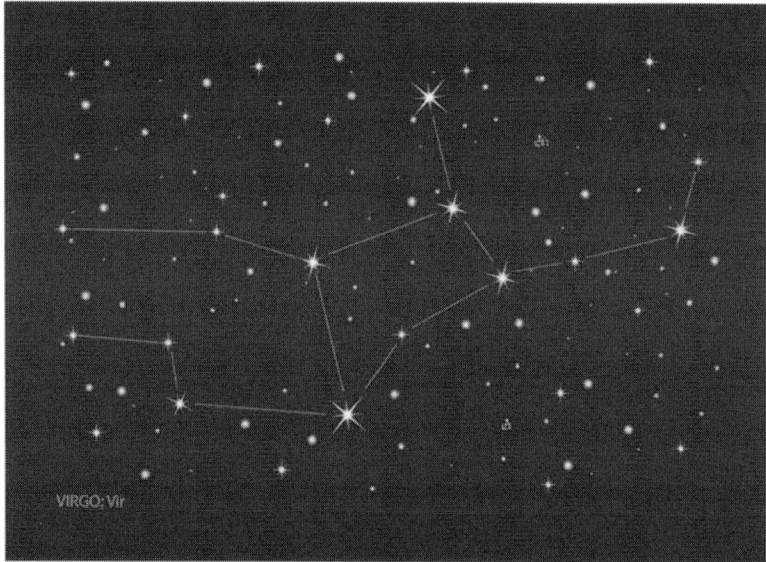

The constellation Virgo takes its name from the Latin for virgin or young maiden.

In ancient Greek mythology, Virgo is associated with the goddess Dike, the goddess of justice. The constellation Libra, which represents the scales of justice, lies next to Virgo.

Virgo lies over the southern hemisphere and is one of the largest constellations in the sky; it is visible between latitudes 80°N and 80°S.

The brightest star in Virgo is **Spica**, the 15th brightest star in the sky and a navigational star which is best seen during nautical twilight in May.

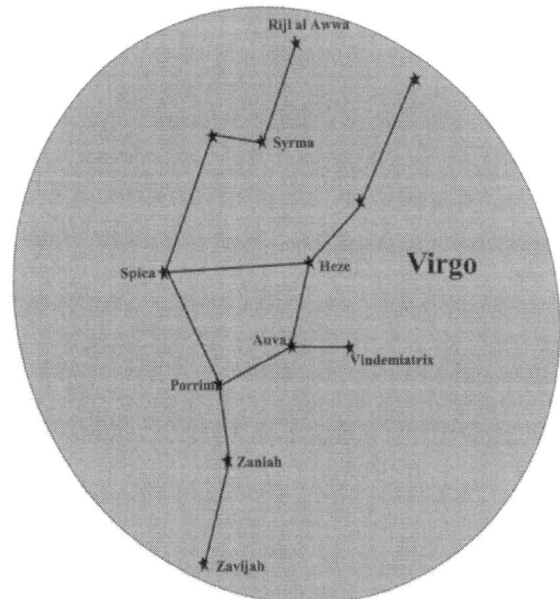

**Finding Virgo.**

As we learned when studying the constellation Boötes, an imaginary curved line from Alkaid in the Great Bear leads to the bright orange star Arcturus, in the constellation Boötes. If, as shown in the following diagram, we continue that curved line by another hand span from Arcturus we will come to the bright bluish-white star Spica in the constellation Virgo.

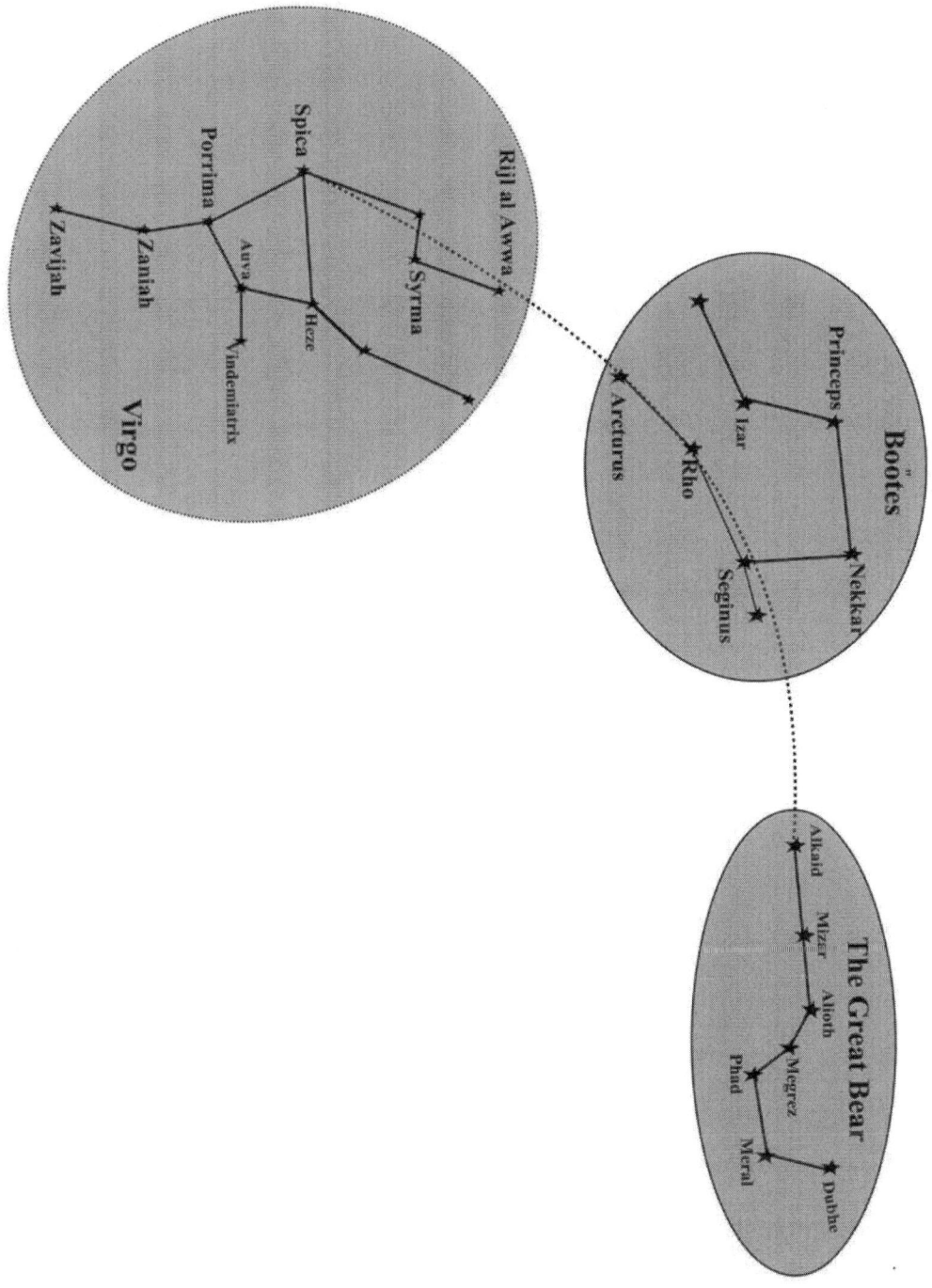

## Libra The Scales

Libra lies in the southern hemisphere between 10° and 30° south and can be seen between latitudes 65°N and 90°S.

The name Libra means 'The Weighing Scales' in Latin and the constellation Libra is depicted as the scales of justice held by the Greek goddess of justice Dike who is represented by the constellation Virgo. In the diagram below, the stars Zuben Elschmali and Zuben Elgenubi mark the scales' balance beam with Elakrab and Brachium representing the weighing pans.

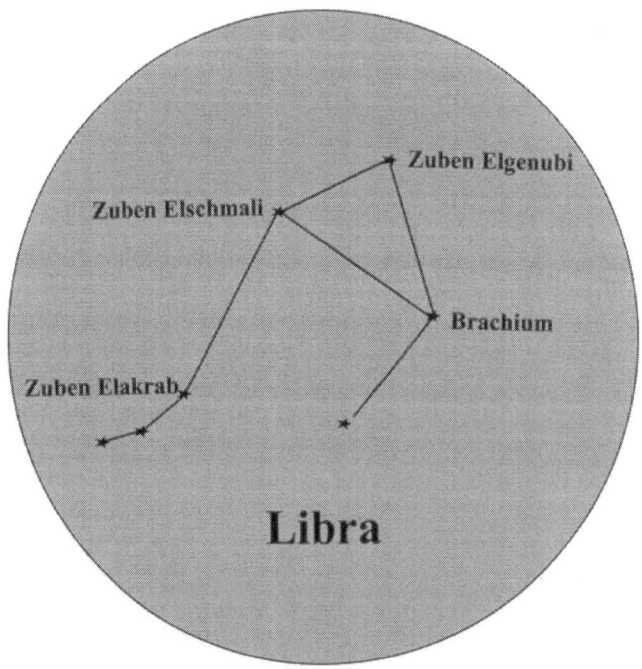

As a matter of interest, the star name Zuben Elgenubi may have been the inspiration for the name 'Obi-Wan Kenobi' of Star Wars fame because of their similar sounds.

Zuben Elschmali is the brightest star in the constellation but it is not considered to be a navigational star. In fact there are no navigational stars in Libra.

**Finding Libra.** This is one of the hardest constellations to find because it has no bright stars. The ancient Greeks considered it to be part of the Scorpio constellation because it was said to represent the claws of the scorpion and this gives us the clue to locating Libra in the sky. The diagram below shows the close proximity of Libra to Scorpius and it is easy to see how the representation described above was imagined.

Zuben Elgenubi is midway between the stars Antares in Scorpius and Spica in Virgo, the brightest stars in their respective constellations. The angular distance from the star Zuben Elgenubi in Libra to the star Geaffias in Scorpio is about 10 degs which is roughly equivalent to the width of the palm of the hand when held at arm's length.

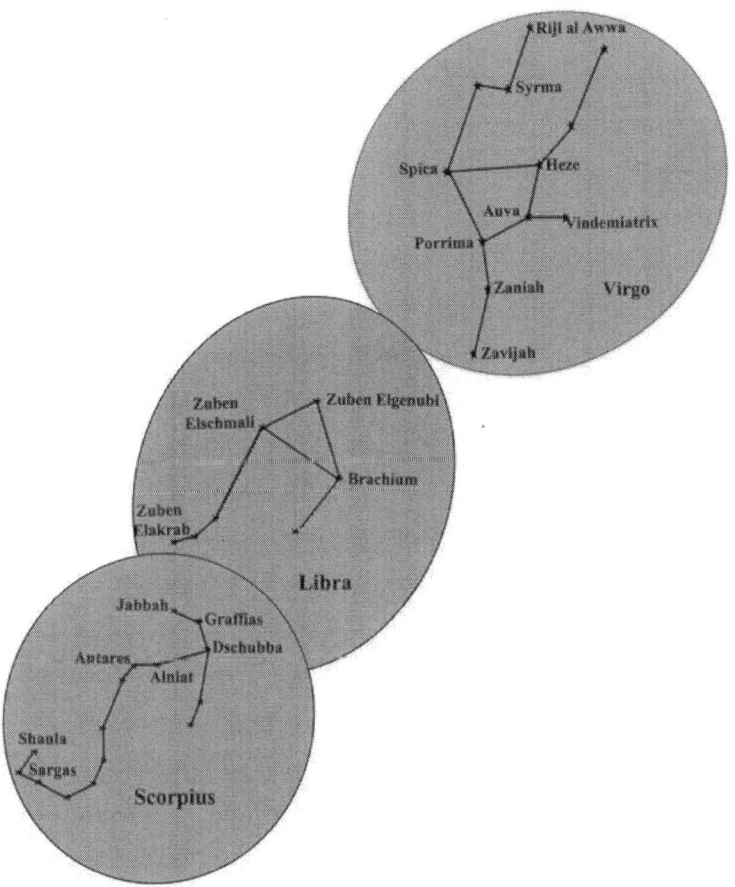

# Cancer The Crab

Cancer is a relatively small constellation in the northern hemisphere and is visible between latitudes 90°N and 60°S.

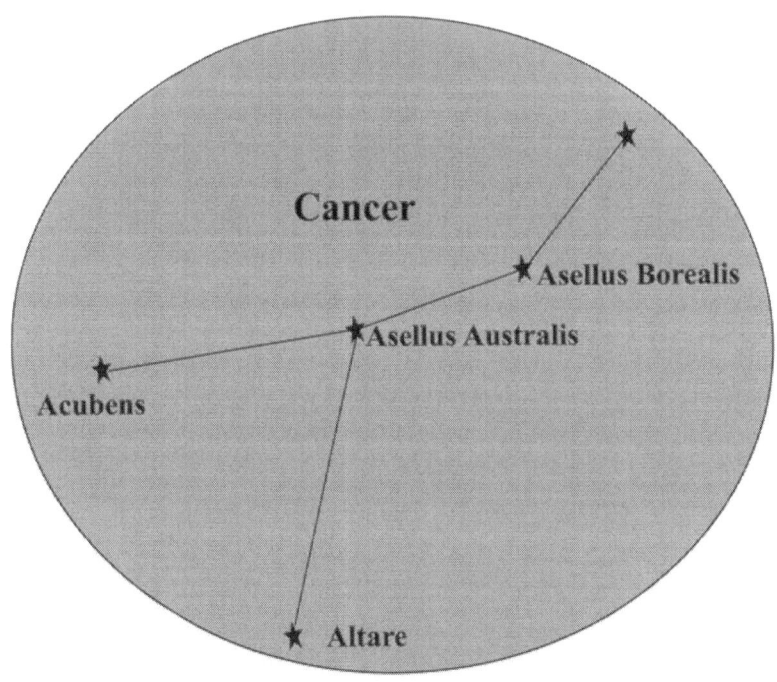

Cancer consists of mainly faint stars, none of which is a navigational star and for this reason, it is not a very useful constellation for astro navigation. However, it does help us in one way: It was stated in the introduction to this chapter that, although this is not intended to be an astrological tour, the signs of the zodiac can be very useful to navigators because the order in which they follow one another can tell us the position of one zodiac constellation in the sky with respect to another. For example, we know that Cancer's position on the ecliptic falls between Leo and Gemini and as the following diagram demonstrates, Cancer can easily be found nestling between those constellations with Gemini lying to the west and Leo to the east (remember that in star maps, east and west are reversed with respect to conventional maps).

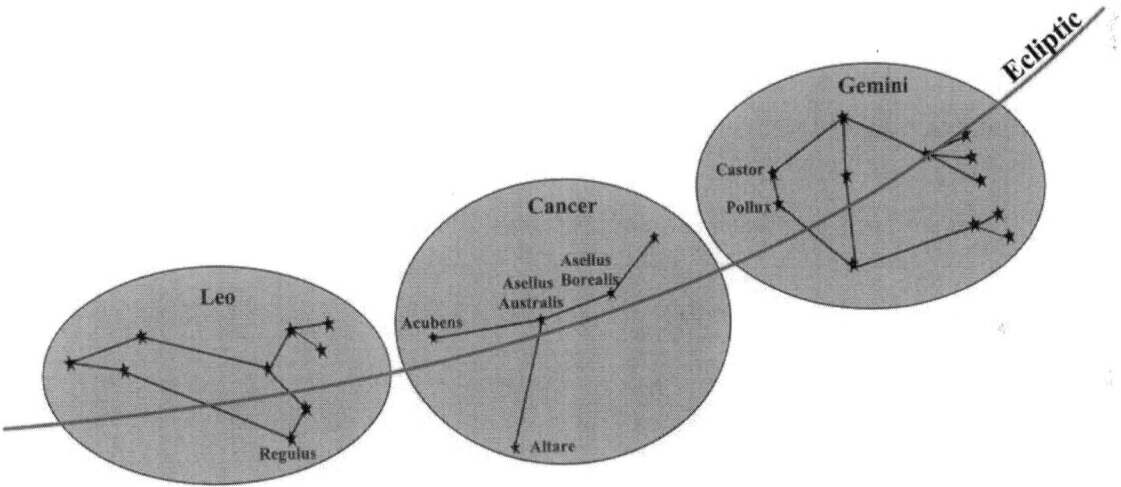

From this, it can be seen that Cancer can be an aid to locating both Gemini and Leo.

These days, the Sun passes through Cancer in late July; however, in the time of Ptolemy, around 2000 years ago, this occurred during the summer solstice when the Sun reached 23.4° N, the northern limit of the ecliptic. The latitude 23.4° N is still called the tropic of Cancer even though the Sun now resides in Taurus at the summer solstice.

In Greek mythology, Cancer is associated with the crab in the story of the Twelve Labours of Heracles. The goddess Hera sent the crab to attack Hercules while he was fighting the Lernaean Hydra but Hecules kicked it all the way to the stars where it formed the constellation Cancer. In another version of the story, Hera placed the crab in the sky in gratitude for its efforts even though it was killed by Hercules. (Heracles is the Roman name for the Greek god Hercules).

# CENTAURUS

Centaurus is one of the largest constellations in the sky and is quite bright with three navigational stars: **Hadar, Menkent** and **Rigel Kentaurus** which are best seen during nautical twilight in May. It lies above the southern hemisphere and is visible between latitudes 25° N and 90°S.

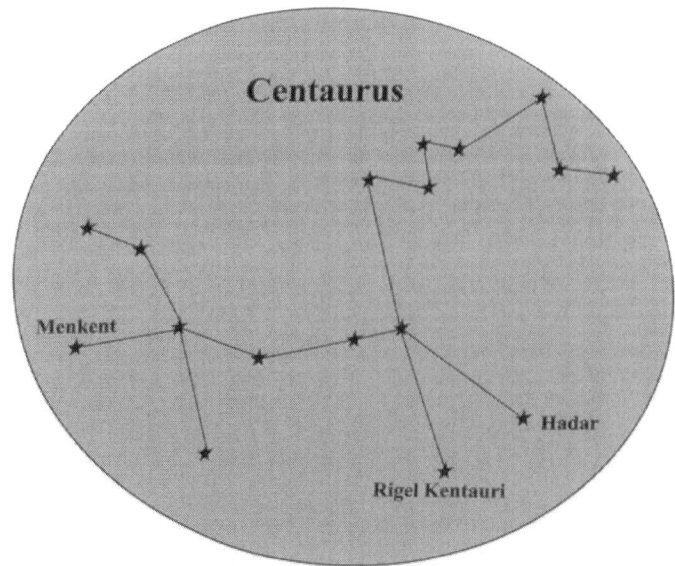

Centaurus is associated with several mythological tales and is depicted as a centaur, half man and half horse. In Ancient Greek mythology, the constellation is associated with Chiron, the centaur who mentored many of the Greek heroes including Theseus, Jason and Heracles.

**Finding Centaurus.** If we take a line from Arcturus in the constellation Boötes to Spica in Virgo and extend that line by another hand-span, it will point to the constellation Centaurus as shown in the diagram below.

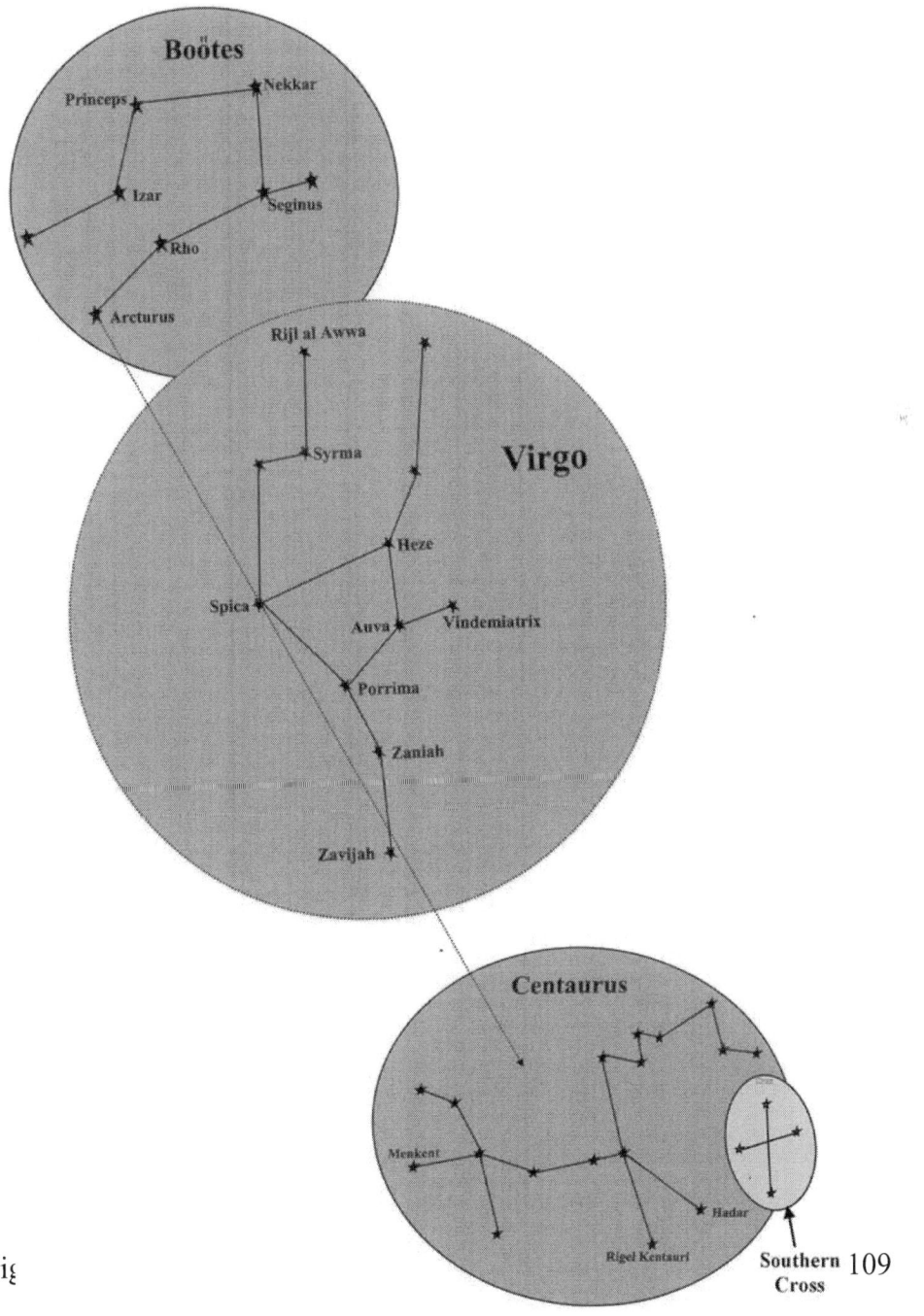

## Crux, The Southern Cross

Crux is one of the best known constellations in the southern hemisphere, and is easily recognizable for the cross-shaped asterism named the 'Southern Cross' which is formed by its four brightest stars.

Crux (Latin for cross), it is one of the smallest constellations in the sky but also one of the brightest which makes it useful for astro navigation. It is not visible north of 20°N in the northern hemisphere but it is circumpolar south of 34°S in the southern hemisphere which means that it never sets below the horizon there.

The cross has four main stars: alpha, beta, gamma and delta which mark the tips of the cross.

**Alpha Crucis** is also known as **Acrux** (a contraction of 'alpha' and 'Crux'). This is the brightest star in the constellation Crux and is a navigational star.

Beta Crucis, also known as Mimosa or Becrux, is the second brightest star in the constellation but is not a navigational star.

**Gamma Crucis** or **Gacrux**, is the third brightest star in Crux and is a navigational star.

Delta Crucis or Palida is the fourth star and has variable levels of brightness making it unsuitable as a navigational star.

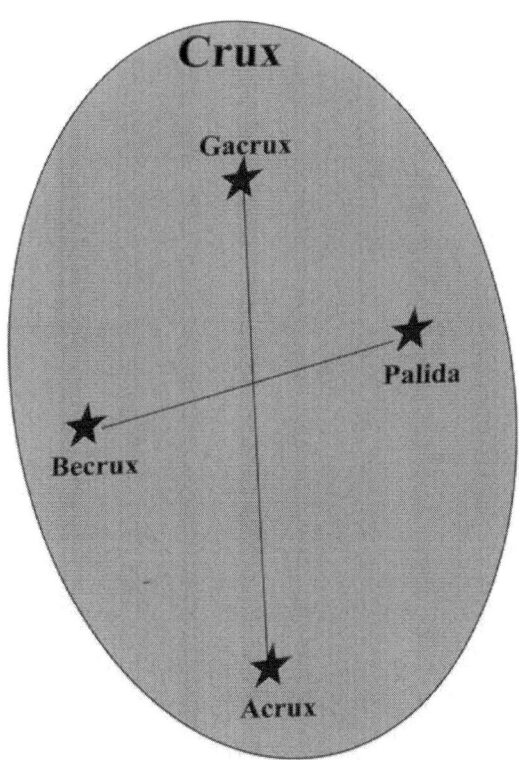

**Mythology.** Crux is associated with the mythology of several cultures. Some Aboriginal cultures saw it the head of the 'Emu in the sky' while others saw it as representing the sky deity Mirrabooka and indeed that is the name that they gave to what we know as the 'Southern Cross'. In New Zealand, the Maori called the cross Te Punga ('the anchor') and in South America, the Incas called it Chakana (the stair).

**Finding Crux.** The constellation Centaurus contains two bright stars which make excellent pointers to help us find the Southern Cross. **The Pointers** as they are known, are **Rigil Kentaurus** and **Hadar** which are both navigational stars. The next diagram shows the relationship between the pointers and the Southern Cross.

The constellation Centaurus appears to envelope Crux and indeed the ancient Greeks considered them to part and parcel of the same constellation.

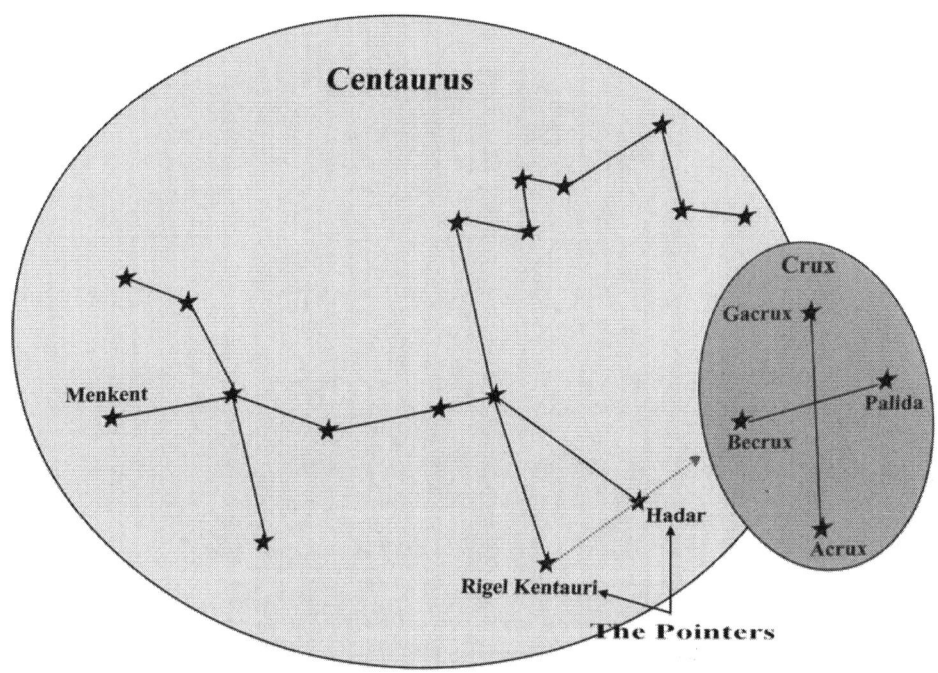

## How to find the direction of south by using the Southern Cross.
There are several methods of doing this but the simplest is as follows: Make an imaginary line between Gacrux and Acrux then extend this line from Acrux (the brightest star) for 4.5 times the length of the Southern Cross, as shown in the diagram below. This will take you to the position of the South Celestial Pole in the sky. From the South Celestial Pole, drop a line down to the horizon. Where this line touches the horizon is the direction of south.

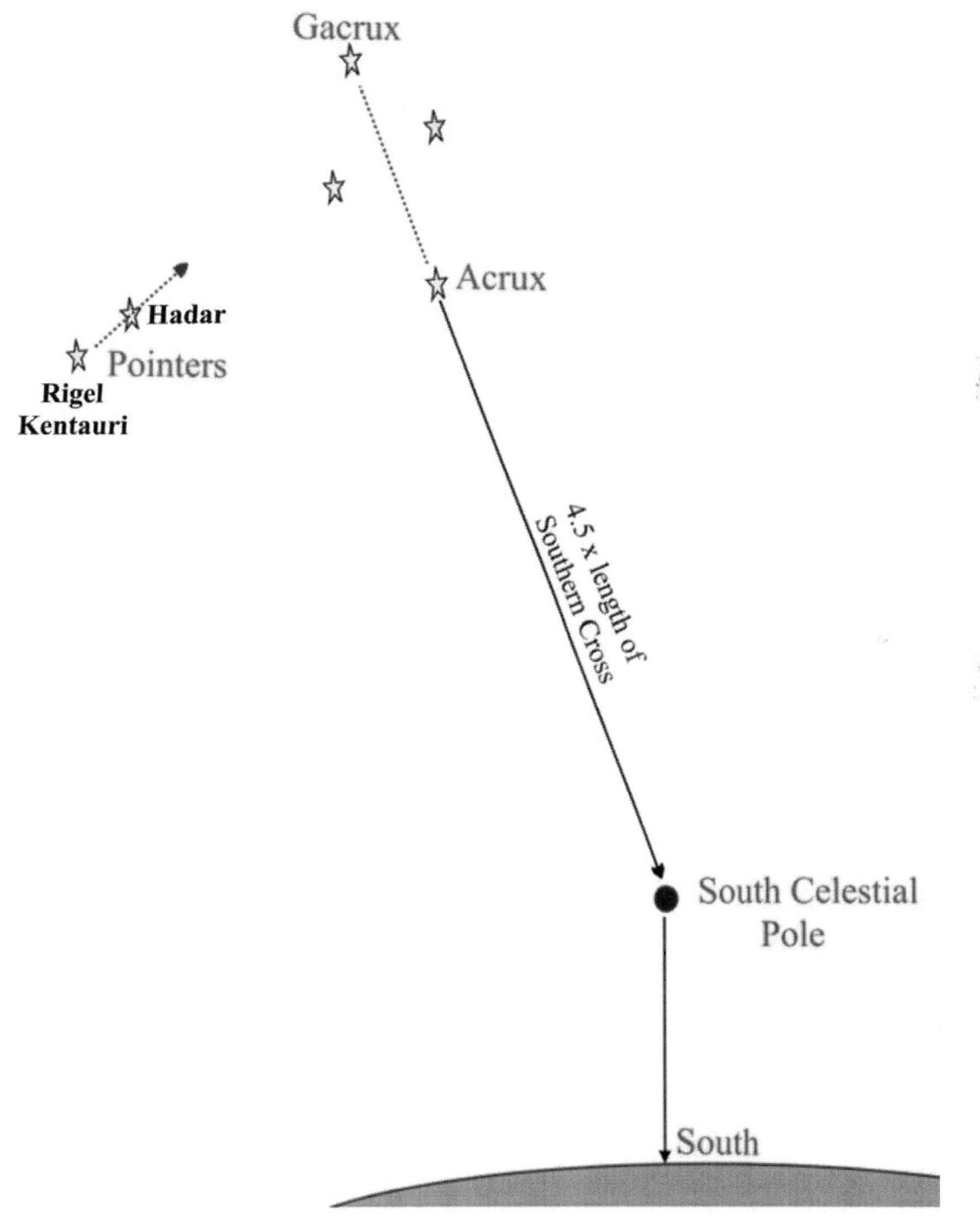

# Chapter 6
# Latitude and Longitude

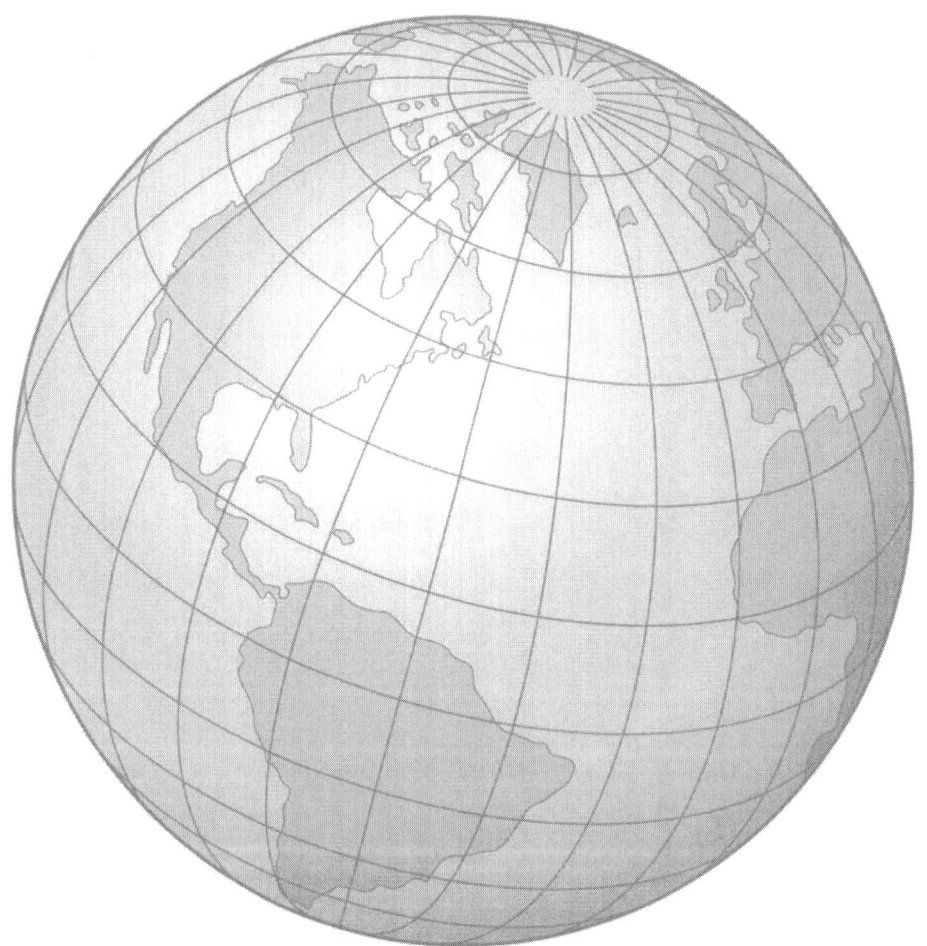

A good starting point for a study of latitude and longitude is to consider the definitions of Great Circles and Small Circles:

**Great Circles.**  A plane of a sphere which passes through the centre of the sphere is called a great circle.  The paths of two great circles of a sphere will intersect at two points 180° apart on the surface of the sphere.

**Small Circles.**  A plane of a sphere which does not pass through the centre of the sphere is called a small circle.  Two small circles of a sphere need not meet.

Now that we have established these definitions, we can begin to understand the concepts of latitude and longitude.

**Longitude.**
In the diagram below,
C represents the centre of the Earth.
N & S represent the North and South poles respectively.

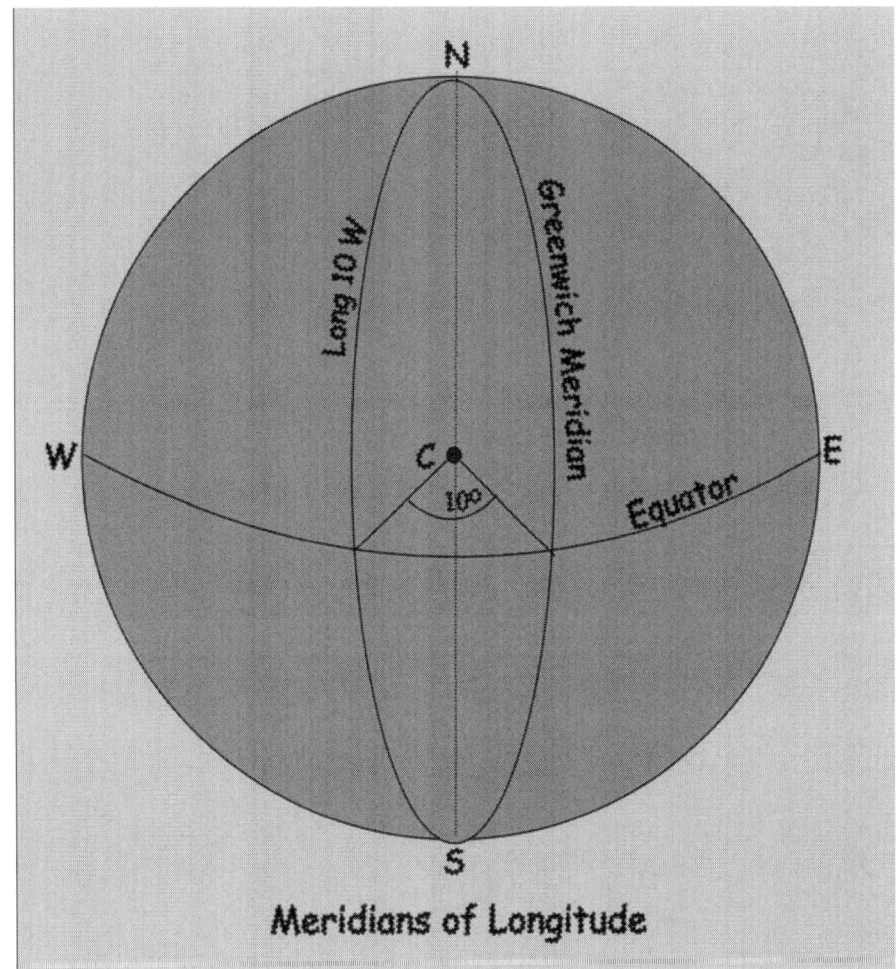
Meridians of Longitude

The great circles that pass through N and S are called **meridians of longitude**. The meridian that passes through Greenwich is used as the base meridian (0°) and all other meridians are described by their angular distance east or west of 0° along the Equator from 1° to 180°. The diagram shows the meridians of longitude 0° (Greenwich Meridian) and longitude 10° West.

**The Equator.** The Equator is an imaginary line around the Earth forming a great circle that is equidistant from the north and south poles and as such, it forms the boundary between the northern and southern hemispheres.

**Latitude.** A section of the Earth's surface made by a plane parallel to the Equator is called a parallel of latitude and by the definitions established above is a small circle.

A parallel of latitude is expressed by its angular distance north or south of the Equator. As illustrated in the diagram below, all points along the parallel of Latitude 10° N have an angular distance of 10° North from the Equator.

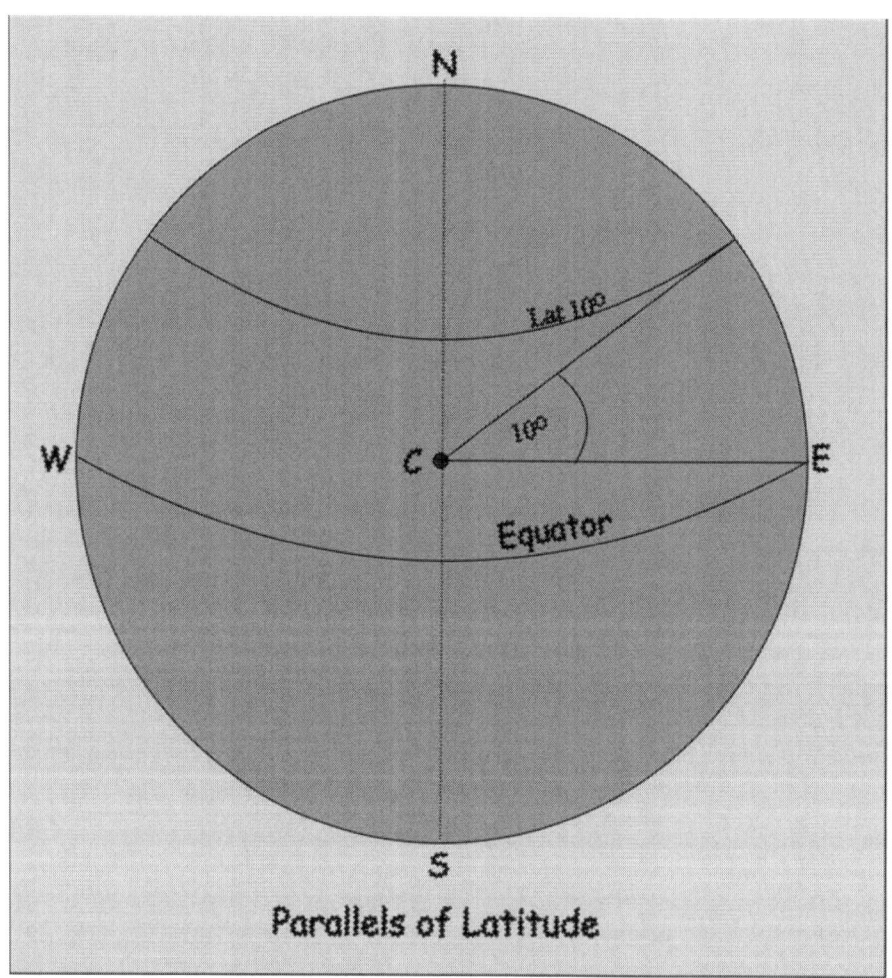

Parallels of Latitude

**Units of Length.**
The standard unit of measurement used for navigation at sea is the nautical mile. However, there are other units of length associated with the word mile and this can lead to confusion for navigators. The most common of these measurements are defined below:

**Statute Mile.** A mile most commonly refers to the statute mile of 5,280 feet (1,760 yards, or 1,609.344 meters). The use of the statute mile as a unit of measurement is largely confined to the United States and the United Kingdom; elsewhere, it has been replaced by the kilometer as a unit of measurement on land.

**Geographical Mile.** The international geographical mile (g.m.) is a unit of length determined by 1 minute of arc along the Earth's equator and is defined as 1855.32 metres.

**Nautical Mile.** The international nautical mile (n.m.) is closely related to the geographical mile and is a unit of length corresponding approximately to one minute of arc along any meridian of longitude. It is defined as exactly 1852 metres.

**Kilometre.** The kilometre (km) is a unit of length in the metric system and is equal to 1000 metres. As stated above, it is used to express distances between geographical places in most countries of the world except for the United Kingdom and the United States where the statute mile is used.

| | Statute Miles | Geographical Miles | Nautical Miles | Metres | Kilometres |
|---|---|---|---|---|---|
| Measurement Conversion Table | | | | | |
| Statute Mile | - | 0.867 | 0.869 | 1609.34 | 1.6 |
| Geographical Mile | 1.153 | - | 1.002 | 1855.32 | 1.86 |
| Nautical Mile | 1.15 | 0.998 | - | 1852 | 1.85 |
| Kilometre | 0.62 | 0.539 | 0.54 | 1000 | - |

**The Circumference of the Earth.** The Earth is not a true sphere and so there can be no single value for its circumference. The rapid rotation of the

Earth about its axis tends to flatten it into an **oblate spheroid** which simply means that the distance between the poles is shorter than the diameter of the equatorial axis giving it the appearance of a squashed ball. The result of this is that the circumference decreases as you move from the equator towards the poles.

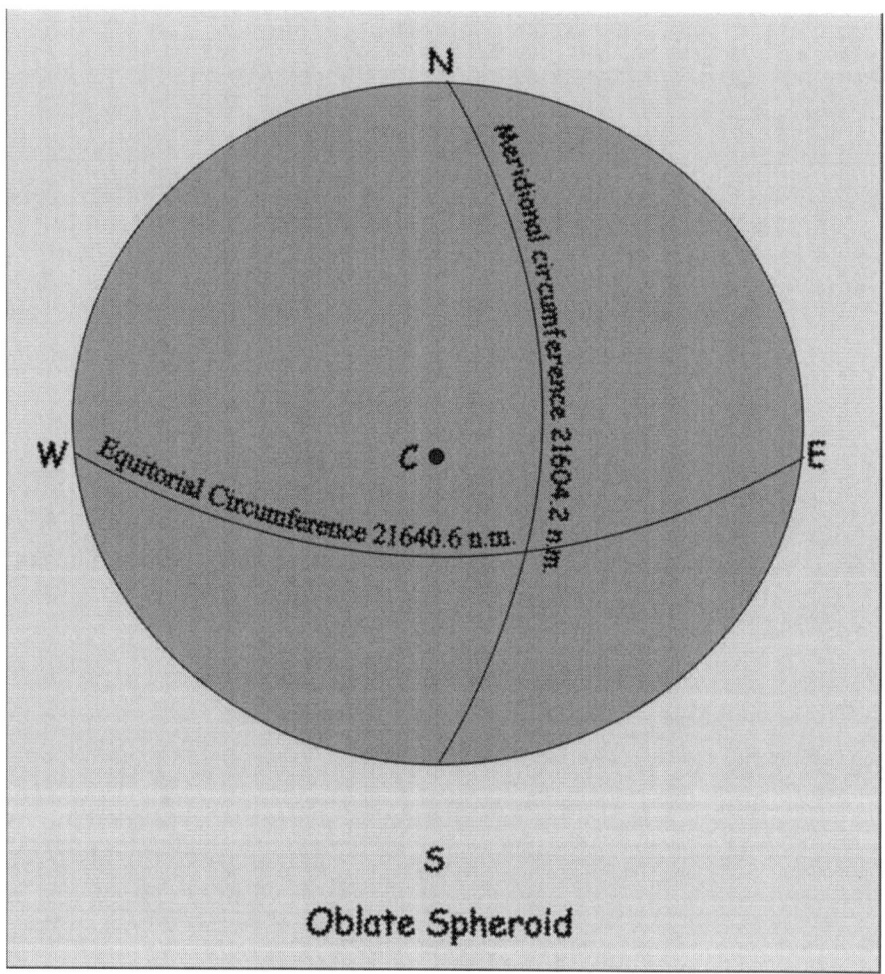

Oblate Spheroid

The three values of the Earth's circumference that we are most interested in are as follows:
Equatorial Circumference = 40075.16 Km.
Polar (or Meridional) Circumference = 40008 Km.
Mean circumference = 40041.58 Km.
For navigational purposes, we need to express measurements, sometimes in terms of nautical miles and sometimes in terms of geographical miles. We

also need to understand various other dimensions of the Earth and be able to convert these from one unit of measurement to another.

The following table lists the most commonly used earth dimensions expressed in terms of different units of measurement.

| Earth Dimensions Table | | | | |
|---|---|---|---|---|
| | Kilometres | Statute Miles | Nautical Miles | Geographical Miles |
| Meridional (Polar) Radius | 6356.8 | 3941.2 | 3432.67 | 3426.3 |
| Equatorial Radius | 6378.1 | 3954.4 | 3444.17 | 3437.8 |
| Mean Radius | 6367.45 | 3947.8 | 3438.42 | 3432.06 |
| Meridional (Polar) Circumference | 40008 | 24805 | 21604.2 | 21564.3 |
| Equatorial Circumference | 40075.16 | 24846.6 | 21640.6 | 21600.5 |
| Mean Circumference | 40041.58 | 24825.8 | 21622.5 | 21582.4 |

**The Relationship Between Longitude and the Nautical Mile.**
As shown in the table, the Earth's equatorial circumference is 21640.6 nautical miles. Since the Equator is a great circle, 1° will subtend an arc on the Earth's surface along the equator of:

$$\frac{21640.6}{360} = 60.113 \approx 60 \text{ nautical miles.}$$

There are 360 meridians of Longitude so it follows that, measuring from the Earth's centre; the angular distance between adjacent meridians at the Equator is 1°. Since 1° subtends an arc of 60 nautical miles, it also follows that the distance between adjacent meridians of longitude at the Equator is 60 n.m.

Returning to the longitude diagram below, the angular difference between longitude 10° West and the Greenwich Meridian is 10°; therefore, the distance between them at the Equator is 10 x 60 = 600 n.m.

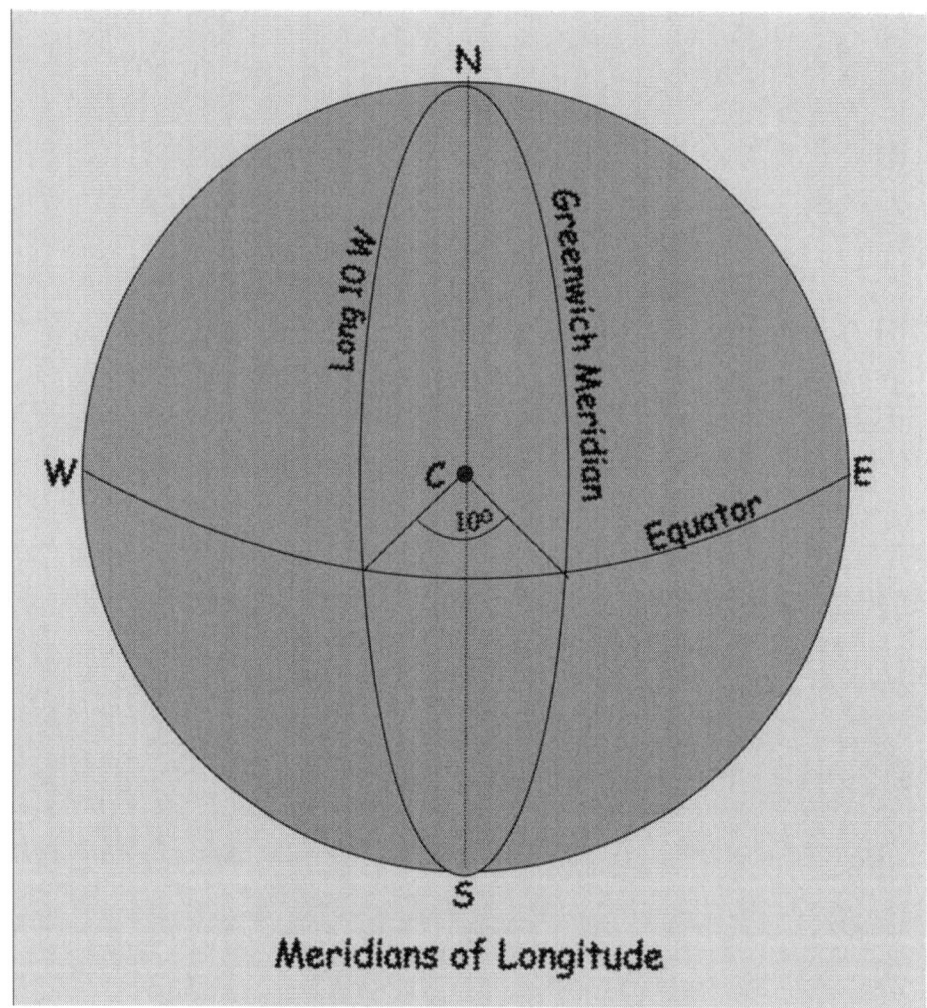

Meridians of Longitude

**Time difference between meridians of longitude.**
We know that the Earth revolves about its axis with respect to the Sun once every 24 hours. In other words, the Sun completes its apparent revolution of 360° in 24 hours. This means that the Sun crosses each of the 360 meridians of longitude once every 24 hours.
So, in 1 hour, the mean sun appears to move 15°,
in 4 minutes, it appears to move 1°,
in 1 minute it appears to move 15',
in 4 seconds it appears to move 1'.
From this, it becomes obvious that there is a direct relationship between arc and time.

## Converting Arc to Time and Vice Versa.

From the above, the local hour angle of the Sun can be expressed in terms of both arc and time and it is important to be able to convert from one to the other. Conversion tables are available in publications such as the Nautical Almanac but if these are not available, it is quite easy and in fact more accurate, to make the conversions by using the following methods:

It is useful to remember the following when making conversions:

$$15° \leftrightarrow 1h$$
$$1° \leftrightarrow 4m$$
$$15' \leftrightarrow 1m$$
$$1' \leftrightarrow 4s$$

### To Convert Arc Into Time.
*Multiply by 4 and divide by 60*
Example.  Convert 65° 30' to time:

```
                              h  m   s
4 x 65° ÷ 60 = 260° ÷ 60 =    4  20  00
4 x 30' ÷ 60 = 120' ÷ 60 =       2   00
              ∴ 65° 30' =     4  22  00
```

### To Convert Time Into Arc.
*Multiply the hours by 15 and divide the minutes and seconds by 4.*
Example.  Convert $12^h$ $32^m$ $15^s$ to arc:

$$12^h = 12 \times 15 = 180°\ 0'\ 0''$$
$$32^m = 32 \div 4 = 8°\ 0'\ 0''$$
$$15^s = 15 \div 4 = 0°\ 3'\ 45''$$
$$\therefore 12^h\ 32^m\ 15^s = 188°\ 3'\ 45''$$

## The Relationship between Latitude and the Nautical Mile.

The Earth's meridional circumference is 40007.86 Km. which equates to 21604.2 nautical miles (n.m.). In other words, an angle of 360° at the Earth's centre subtends an arc of 21604.2 n.m. on the surface of a meridian of Longitude which is by definition a great circle.

From the above, it follows that:
1° measured along a meridian of longitude, will subtend an arc of:
$$\frac{21604.2}{360} = 60.012 \approx 60 \text{ n.m.}$$
and 1' will subtend an arc of:

$$\frac{60}{60} = 1 \text{ n.m}$$

The next diagram shows the parallel of latitude 1° North.
As calculated above, an angle of 1° at the Earth's centre will subtend an arc of 60 n.m. along a meridian of longitude. Therefore, any point on the parallel of latitude 1° North will have an angular distance of 60 n.m. north of the Equator.

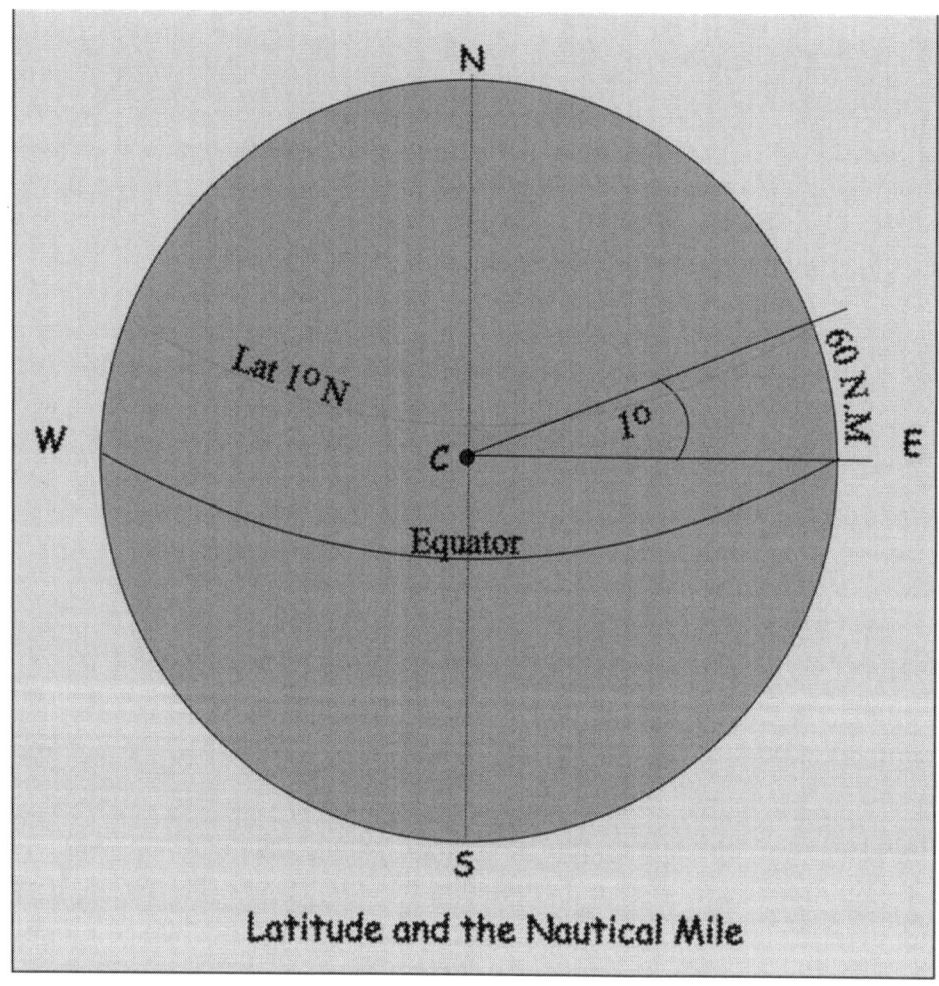

**Latitude and the Nautical Mile**

**Distance Between Parallels of Latitude.**
The angular distance between latitudes 62° N and 25° N is 37°. Therefore, the distance between these parallels, measuring due north or south along any

meridian of longitude, will be: 37 x 60 = 2220 nautical miles as shown in the following diagram.

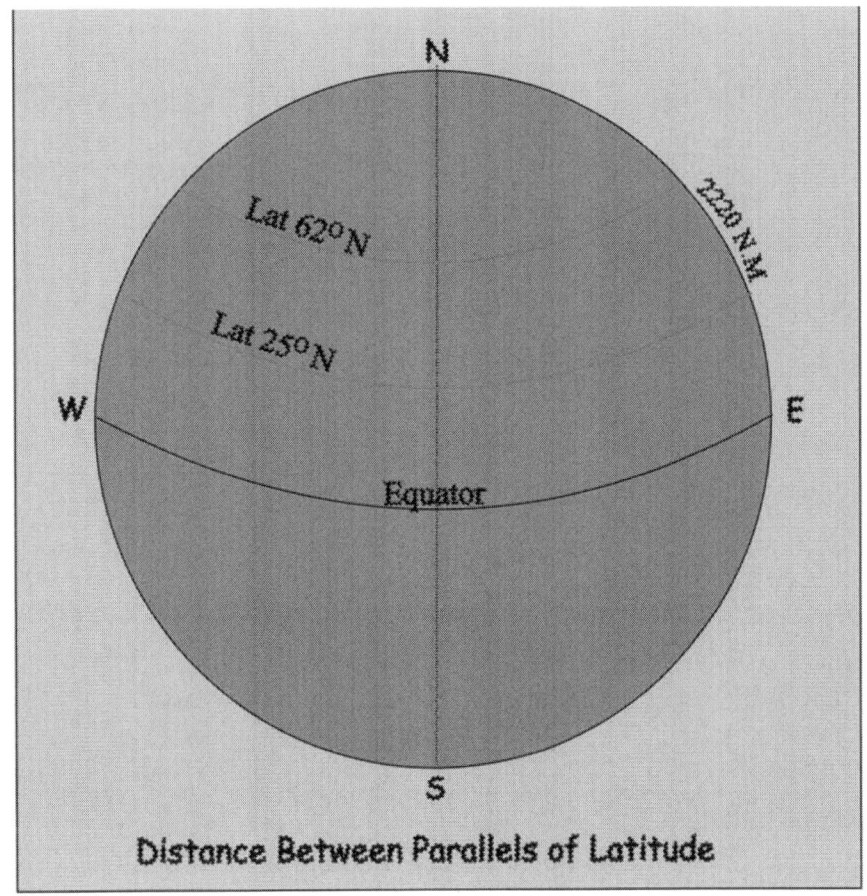

**Distance Between Parallels of Latitude**

# Chapter 7
# Time

**Definitions of Time.**
The apparent movement of the Sun provides the basis for our definitions of time.

**The True Sun.**
Life on Earth is governed by the movement of the True Sun, that is, the sun we see in the sky and not by the theoretical Mean Sun. The units of time in everyday use are defined in terms of the True Sun and are known as **Apparent Solar Time.**

**The Sun as a Time-Keeper.**
Because the Earth's orbital motion is not uniform, there are corresponding variations in the apparent speed of the Sun along the ecliptic. For this reason, the hour angle of the True Sun does not increase at a uniform rate

and therefore does not give an accurate measurement of time. To overcome this problem without losing the connection with the True Sun, the Mean Sun, as defined below, is used.

### The Mean Sun.

The Mean Sun is an imaginary body which is assumed to move in the celestial equator at a uniform speed round the Earth and to complete one revolution in the time taken by the True Sun to complete one revolution of the ecliptic. Although the imaginary Mean Sun gives us an accurate measurement of time, it presents the navigator with a problem. When fixing his position by an observation of the Sun, he measures the altitude of the True Sun which keeps apparent solar time. However, he notes the time of the observation from a deck watch that keeps **mean solar time**. To enable us to connect mean solar time with apparent solar time, we have the Equation of Time which is explained below.

### The Equation of Time.

The True Sun moves with varying speed along the path of the ecliptic while the Mean Sun moves with constant speed along the path of the celestial equator. Because of this, their hour angles do not keep in step with each other and therefore we must be able to connect mean solar time with apparent solar time. The equation of time is designed to provide this connection and is defined as the excess of mean time over apparent time; that is:

### Equation of Time = Local Hour Angle of Mean Sun - Local Hour Angle of True Sun

This can be abbreviated to **LHAMS - LHATS**

At certain times of the year, LHATS will be greater than LHAMS and at other times, it will be less with the values ranging from +15 to -15.
From 15th June to 31st August and from 25th December to 14th April, LHAMS is less than LHATS so the values of the equation of time are negative.
From 15th April to 14th June and from 1st September to 24th December, LHATS is less than LHAMS and so the values are positive.

The above is demonstrated in the following diagrams.

Notes pertaining to the diagrams:
1. Local Hour Angle (LHA) is measured westwards from the meridian of the observer).
2. TS: True Sun.   LHATS: LHA of True Sun.
3. MS: Mean Sun.   LHAMS: LHA of Mean Sun.
4. Z: Zenith of observer.
5. Symbol for First Point of Aries :  ♈

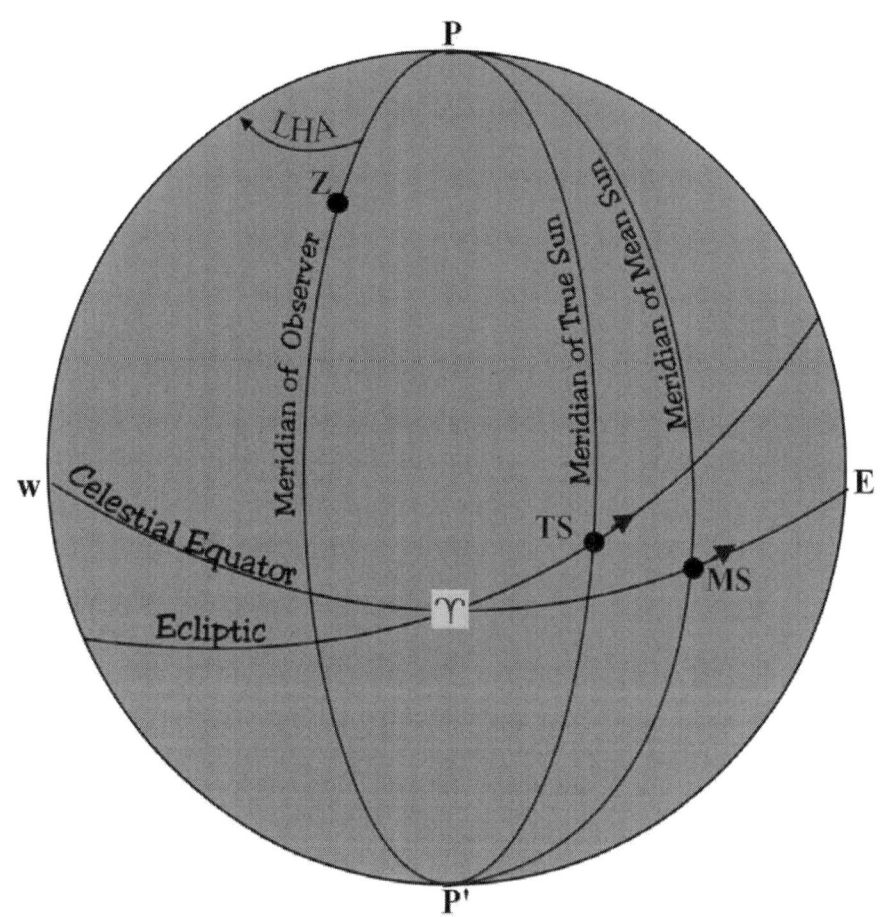

**Equation of Time Negative**
(LHAMS < LHATS)

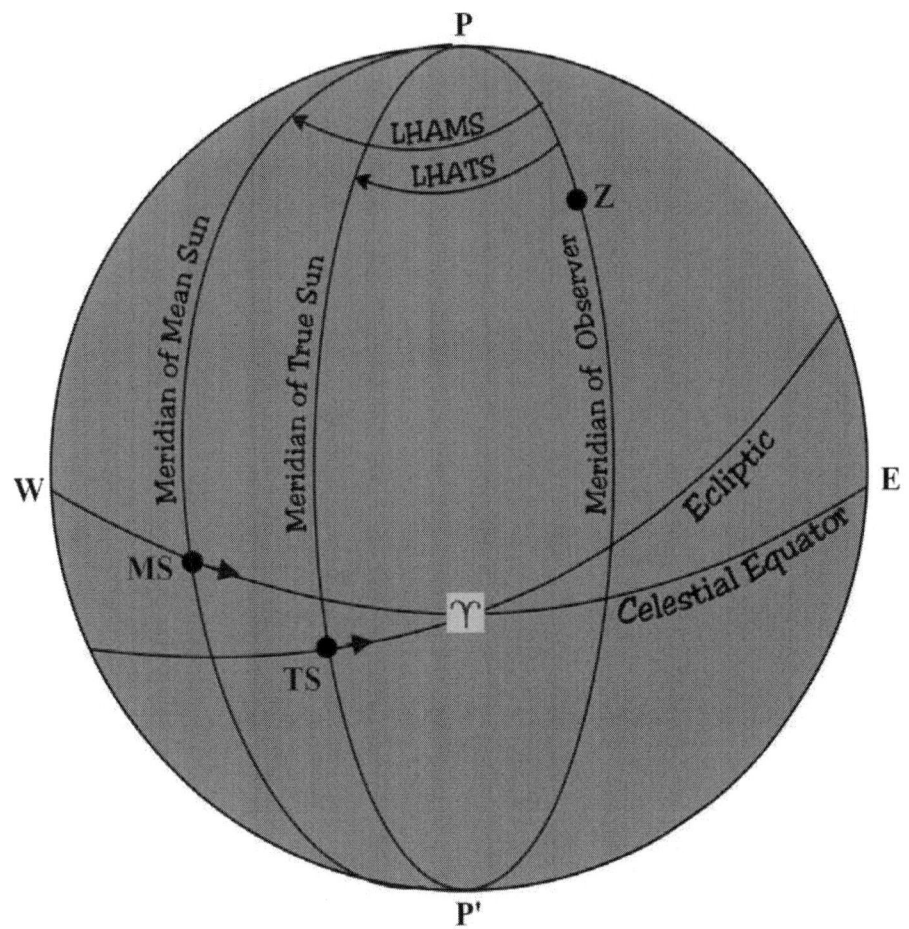

**Equation of Time Positive**
(LHAMS > LHATS)

♈ The First Point of Aries is the point at which the True Sun crosses the celestial equator as it moves northwards along the ecliptic. This event occurs at the Vernal Equinox and when it does, the equation of time becomes zero. We know that the True Sun crosses the celestial equator again at the Autumnal Equinox and it would seem that these are the only two occasions when the equation of time becomes zero. However, things are never that simple; other factors come into play which cause the equation of time to become zero and change sign on four occasions during the course of a year. These occasions fall on approximately the following dates: 15th April, 14th June, 1st September and 24th December.

In brief, the other factors that affect the equation of time include the following:

a. **Obliquity**. This is caused by the Earth's tilt which leads to changes in the angle between the planes of the equator and the ecliptic as the Earth orbits around the Sun.

b. **Eccentricity**. This is caused by the attraction of the Sun, Moon and planets on the Earth which results in the unequal motion of the Earth as it travels around its elliptical orbit.

The chart below shows that the equation of time becomes zero and changes sign on four occasions during the course of a year.

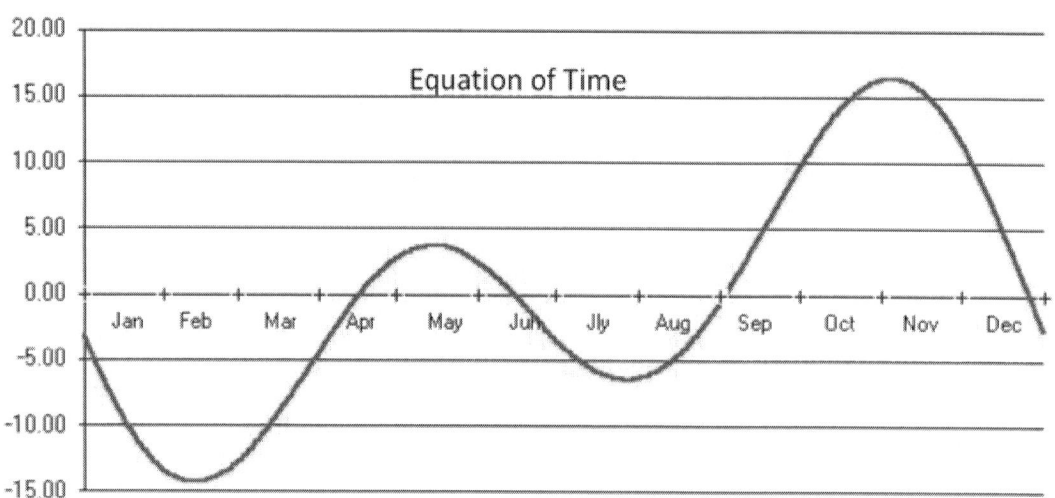

**Summary.**
- The equation of time can be either positive or negative depending on the time of the year.
- The values range from approximately +15 to -15 mins.
- The values are positive from 15th April to 14th June and from 1st September to 24th December.
- The values are negative from 15th June to 31st August and from 25th December to 14th April.

To simplify matters for the navigator, the value of the equation of time is tabulated against GMT in the Nautical Almanac for every 12 hours.

### Example.
An observation of the Sun is made at 11h 05m 22s GMT on 26 August 2009. What is the apparent solar time?

The daily page for 26 Aug. shows that the Eqn. of Time on that date is -01m 55s.

Eqn. of Time = mean solar time - apparent solar time.

∴ apparent solar time = mean solar time - Eqn. of Time.
= 11h 05m 22s - (-)01m 55s

∴ apparent solar time = 11h 07m 17s

**The Civil Day** is the day that is defined for use for human activity rather than for astronomical purposes. It begins at midnight when the Local Hour Angle of the Mean Sun is 12 hours or 180° and ends the following midnight. It is divided into two periods of 12 hours each. The first period consists of the 12 hours from midnight to noon and is denoted by the abbreviation a.m. (ante meridian). The second period is the 12 hours from noon to midnight and is denoted by p.m. (post meridian).

**The Astronomical Day** consists of one period of 24 hours instead of two periods of 12 hours. This system is more convenient for tabulation purposes since times can be written as 4 figures and dispenses with the need for the abbreviations a.m. and p.m. For example, 11.45 a.m. is simply written as 1145 and 7.32 p.m. is written as 1932.

**The Mean Solar Day** is the time taken for the Mean Sun to make one complete circuit of the Earth. In other words, it is the time taken for the Mean Sun to transit all 360 meridians of longitude. The time system based on the Mean Solar Day is known as **Mean Solar Time.**

**Mean Noon** occurs when the meridian of the Mean Sun coincides with the meridian of a place.

**Apparent Noon** is when the True Sun is on an observer's meridian of longitude. It is when the Sun reaches its greatest altitude above the observer's horizon. In other words it is when the Sun is at its **zenith**.

Local Hour Angle (LHA). As explained in chapter 2, the Local Hour Angle of the Sun is the angle between the meridian of the observer and the meridian of the Sun's Geographical Position. LHA is measured westwards from the observer's meridian and can be expressed as angular distance or as time.

**Converting Arc to Time and Vice Versa.** As previously discussed, due to the Earth's rotation, the Sun appears to move through 360° of longitude in 1 day, 15° in 1 hour and 15′ in 1 minute of time and therefore, the local hour angle of the Sun can be expressed in terms of both arc and time.

It is useful to remember the following when making conversions:

$$15° \leftrightarrow 1h$$
$$1° \leftrightarrow 4m$$
$$15' \leftrightarrow 1m$$
$$1' \leftrightarrow 4s$$

Conversion tables are available in publications such as the Nautical Almanac but if these are not available, it is quite easy and in fact more accurate, to make the conversions by using the following methods:

**To Convert Arc Into Time.**
*Multiply by 4 and divide by 60*
Example. Convert 65° 30′ to time:

```
                                    h  m  s
4 × 65° ÷ 60 = 260° ÷ 60 =          4 20 00
4 × 30′ ÷ 60 = 120′ ÷ 60 =            2 00
            ∴ 65° 30′ =             4 22 00
```

**To Convert Time Into Arc.**
*Multiply the hours by 15 and divide the minutes and seconds by 4.*
Example. Convert $12^h\ 32^m\ 15^s$ to arc:

$$12^h = 12 \times 15 = 180°\ 0'\ 0''$$
$$32^m = 32 \div 4 = 8°\ 0'\ 0''$$
$$15^s = 15 \div 4 = 0°\ 3'\ 45''$$
$$\therefore 12^h\ 32^m\ 15^s = 188°\ 3'\ 45''$$

**Local Mean Time (LMT)** is the local hour angle of the Mean Sun measured westwards from the meridian of a certain place and expressed in terms of time instead of arc.

In the following diagram, imagine we are looking down on the North Pole from space.

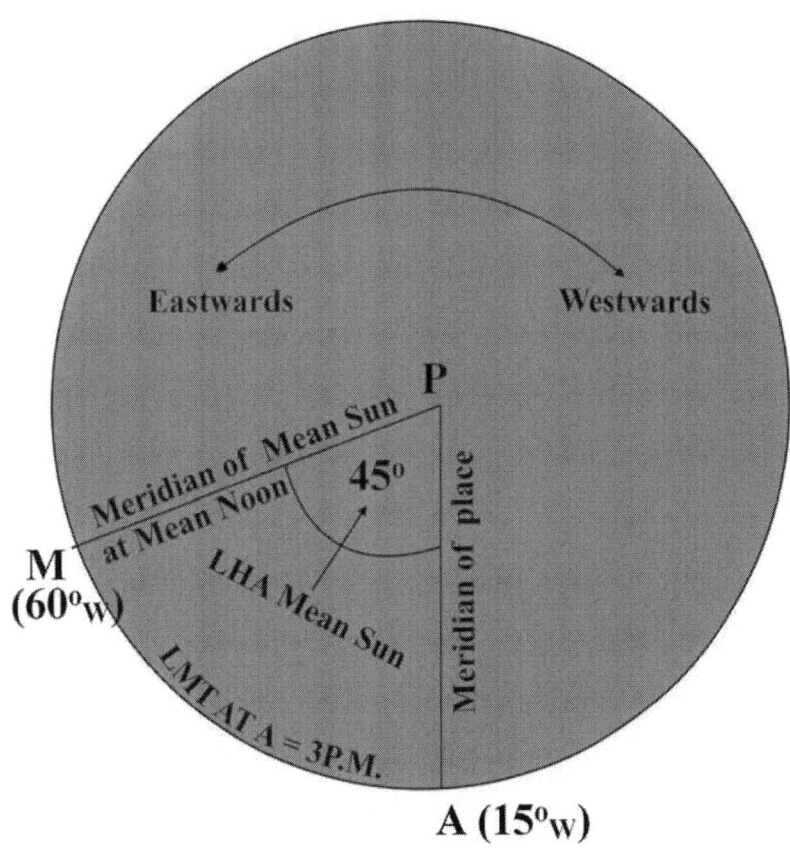

## Local Mean Time

Point A is a place on the Earth's surface and AP represents part of the meridian of longitude of that place.

MP is part of the meridian of longitude on which, for a brief instant, the Mean Sun lies and M is a point on the Earth's surface which lies on that meridian.

Point P is the North Pole

Angle APM is the Local Hour Angle of the Mean Sun.

Suppose the meridian of the Mean Sun is 60°W and the longitude of point A is 15°W, then the LHA of the Mean Sun will be 45°. Since the Mean Sun moves 15° westwards in 1 hour, the time difference between point A and point M will be 3 hours. Because point A is 45° to the east of the Mean Sun's meridian, it must be 3 hours after Mean Noon and so the local mean time at point A will be 3 hours after noon or 3p.m.

**Note.** The difference between the LHA of the Mean Sun and the LHA of the True Sun is that the former is measured from the Mean Sun's meridian and the latter is measured from the meridian of the GP of the True Sun.

**Greenwich Mean Time (GMT)** is the local mean time anywhere on the meridian of Greenwich. In other words it is the Local Hour Angle of the Mean Sun on the meridian of Greenwich.

Since the Greenwich meridian is used as the base meridian from which the longitude of all places on Earth are identified, it provides the link between the LMT of a place and the LMT at Greenwich (or GMT).

Imagine that we are looking down on the North Pole in the diagram below.

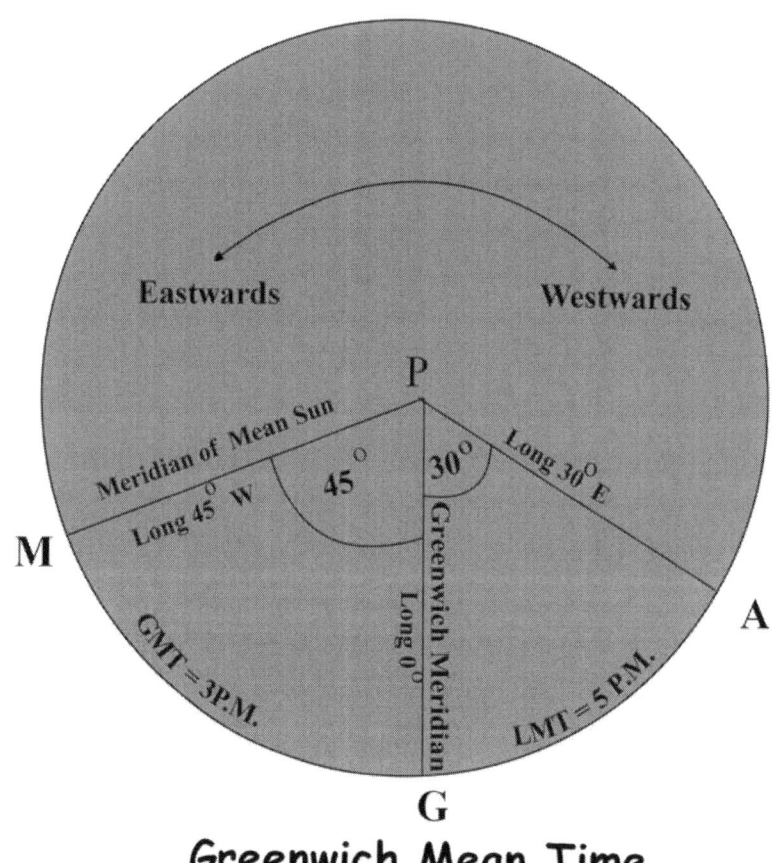

**Greenwich Mean Time**

P represents the North Pole
G represents the position of Greenwich on the Earth's surface.
GP represents part of the Greenwich Meridian (0°).
MP represents part of the meridian of longitude 45°W and M is a point on that meridian.

AP represents part of the meridian of longitude 30° E and A is a point on that meridian.

The meridian of the Mean Sun, for a very brief instant, coincides with the meridian 45°W and so, at that instant, the Local Mean Time at point M is noon.
At the same instant, the Local Hour Angle of the Mean Sun at Greenwich is 45°. Therefore, the LMT at Greenwich must be 3p.m. since the time difference for 45° is 3 hours and Greenwich is to the East of M.
It follows that the Greenwich Mean Time must also be 3p.m. (since GMT is equal to the LMT at Greenwich).

The LMT at point A must be 2 hours after GMT (since the time difference for 30° is 2 hours and A is to the East of Greenwich).
Therefore, the LMT at point A must be 5p.m.

The following **Aide Memoire** is useful when trying to remember whether to add or subtract the time difference:
>**Long West, GMT Best.**
>**Long East, GMT Least.**

### Examples.
1. When it is noon at Greenwich, the LMT at a place on longitude 15°W will be 11 a.m. (since the time difference for 15° is 1 hour and the longitude is west).
2. If Greenwich Mean Time is 4 p.m., the LMT at a place 45°E will be 7 p.m.
3. If the LMT at longitude 75°W is 0130, what is GMT? The time difference for 75° is 5 hours and since the longitude is West, GMT must be +5 hours. Therefore, GMT is 0630.
4. If the time is 2130 GMT, what is the LMT at longitude 150° 15'E.? The time difference is 10 hours, 1 minute. Since the longitude is East, LMT must be greater than GMT. Therefore, LMT is [(2130 + 1001) − 2400] = 0731 the next day.

**Universal Time (UT).** The term Universal Time was adopted internationally in 1928 as a more precise term than Greenwich Mean Time, because GMT can refer to either an astronomical day or a civil day.

**Coordinated Universal Time**, abbreviated as **UTC**, is the primary time standard by which the world regulates clocks and time. It is, within about 1 second of mean solar time at 0° longitude. It is one of several closely related successors to Greenwich Mean Time (GMT). For most purposes, UTC is considered to be interchangeable with GMT, but GMT is no longer precisely defined by the scientific community.
However, the term Greenwich Mean Time persists in common usage to this day and is generally considered to be synonymous with the term Universal Time. It should be noted that the Nautical Almanac and other tables of astronomical data usually refer to UT instead of GMT.

**Standard Time.** For each place on Earth to keep its own local mean time would obviously cause a great deal of confusion and difficulty. For this reason, Standard Time is introduced so that places in the same locality can keep the same time. The time chosen is usually based on a convenient meridian running through the area and the meridian chosen usually differs from the Greenwich meridian by a whole number of hours. Some large countries, such as the U.S.A. have several different standard times.

**Zone Time (ZT).** It would be impossible for a ship at sea to keep to the time of its longitude because (unless it is travelling due north or south) the longitude will be constantly changing. For this reason, the sea areas of the Earth are divided into time zones which are 15° (or 1 hour) apart. The central meridian of each zone is an exact number of hours distant from the Greenwich meridian so that zone time differs from GMT by multiples of 1 hour. The time kept in each zone is the time of its central meridian and is plus or minus GMT depending on the zone's position east or west of Greenwich.

**Time Zone System.** The Earth is divided into 24 zones of 15° of longitude each, with the centre of the system being the Greenwich meridian. Therefore, the centre zone (zone 0) lies between 7.5° W. and 7.5° E. The zones lying to the West of Greenwich are numbered from +1 to +12 and those to the East from -1 to 12. The 12$^{th}$ zone is divided by the meridian

180° which is known as the International Date Line. The two halves of the 12th zone are marked + or – depending on which side of the date line they lie.

The following table shows the centre meridian of each of the 24 time zones. It should be remembered that each zone is 15° wide and extends 7.5° either side of its centre meridian. Thus, since the centre meridian of zone +4 is 60° W, its limits are 52.5° W to 67.5°W.

| 0 | +1 | +2 | +3 | +4 | +5 | +6 | +7 | +8 | +9 | +10 | +11 | 12 |
|---|---|---|---|---|---|---|---|---|---|---|---|---|
| 0 | 15w | 30w | 45w | 60w | 75w | 90w | 105w | 120w | 135w | 150w | 165w | 180 |

| 0 | -1 | -2 | -3 | -4 | -5 | -6 | -7 | -8 | -9 | -10 | -11 | 12 |
|---|---|---|---|---|---|---|---|---|---|---|---|---|
| 0 | 15E | 30E | 45E | 60E | 75E | 90E | 105E | 120E | 135E | 150E | 165E | 180 |

**To Convert Zone Time to GMT.** The time kept in zone 0 is GMT. To convert any other zone time to GMT, simply apply the sign and number of the zone to the zone time. For example, if it is 1800 in zone +5, GMT will be 2300.

**The Greenwich Date (G.D.)** is obtained by converting the zone time and date to GMT
Examples:
ZT 0230(-8), 15 Apr = GD 1830, 14 Apr.
ZT 1915(+10), 24 Dec = GD 0515, 25 Dec.

**Practical Units of Time.** The following units of time are defined in terms of the True Sun and although they are not all used in everyday life, they are important for navigational purposes and therefore need to be defined clearly.

**The Year** is the time taken for the Earth to complete one orbit of the Sun.

**The Day** is the time taken for the Earth to complete one revolution about its own axis. In other words, it is the time taken for the True Sun to make an apparent transit of all 360 meridians of longitude.

**The Hour.** The day is divided into 24 hours and in 1 hour, the Sun will make an apparent transit of 15° of longitude.

**The Minute.**   The hour is divided into 60 minutes and in 1 minute of time, the Sun will transit 15 minutes (15') of longitude or, in other words, a $\frac{1}{4}$ of 1° of longitude.

**The Second.**   The minute is divided into 60 seconds and in 1 second of time, the Sun will transit 15 seconds (15") of longitude or, in other words, a $\frac{1}{4}$ of 1' of longitude.

**Twilight** is the time between dawn and sunrise and between sunset and dusk when the Sun is just below the horizon.  It is so called because scattered sunlight in the upper atmosphere illuminates the lower atmosphere so that the surface of the Earth is neither completely lit nor completely dark.

**Civil Twilight.**  Morning civil twilight begins when the geometric center of the Sun is 6° below the horizon and ends at sunrise.  Evening civil twilight begins at sunset and ends when the geometric center of the Sun reaches 6° below the horizon.  During civil twilight, the horizon is clearly visible and the brightest stars as well as the navigational planets Venus, Mars, Jupiter and Saturn can be seen.

**Nautical twilight** is the time when the center of the Sun is between 6° and 12° below the horizon. Thus, morning nautical twilight ends when morning civil twilight begins and evening nautical twilight begins when evening civil twilight ends. Nautical twilight is so named because it is when navigators are able to take reliable sights of stars and planets using a visible horizon for reference.

**Star and Planet Observations.** In general, the optimum conditions for taking observations of stars and planets occur during the times of civil twilight and nautical twilight when it is likely to be light enough for the horizon to be seen yet dark enough for those celestial bodies to be visible.

The difficulty in using stars and planets for Astro navigation is that it must be dark enough for us to be able to see them yet light enough for us to measure their altitudes above the horizon. This phenomenon generally occurs during civil and nautical twilight'.

**The Sun.** As long as we have clear skies, we can use the Sun during daylight hours for 'position fixing'; however, refraction is greatest when the Sun is close to the horizon and this will make altitude measurement difficult and unreliable at these times.

**The Moon.** The moon can be used for position fixing when it is above the horizon as long as the horizon can be seen. There are three situations when this can occur: during nautical and civil twilight; at night as long as the moon creates sufficient light for the horizon to be seen and during daylight hours, particularly during the morning and late afternoon.

# Chapter 8
# Using Celestial Bodies For Navigation

As explained in chapter 2, our aim in astro navigation is to determine our position in relation to the geographical position (GP) of a celestial body by calculating the altitude and azimuth of that body.

However; this is not as straightforward as we might suppose. For example, suppose we measure the altitude of the Sun from a ship at sea and find it to be 35°; what does this tell us? All that we know is that the ship lies somewhere on the circumference of a circle centered at the geographical position of the Sun. Such a circle is known as a **'position circle'** since our position is known to lie somewhere on its circumference.

The diagram below shows that, at any point on the circumference of the circle, the Sun's altitude will be 35° and that our distance from the GP will be equal to the radius of that circle. The problem is to establish at which precise point on the position circle the ship lays.

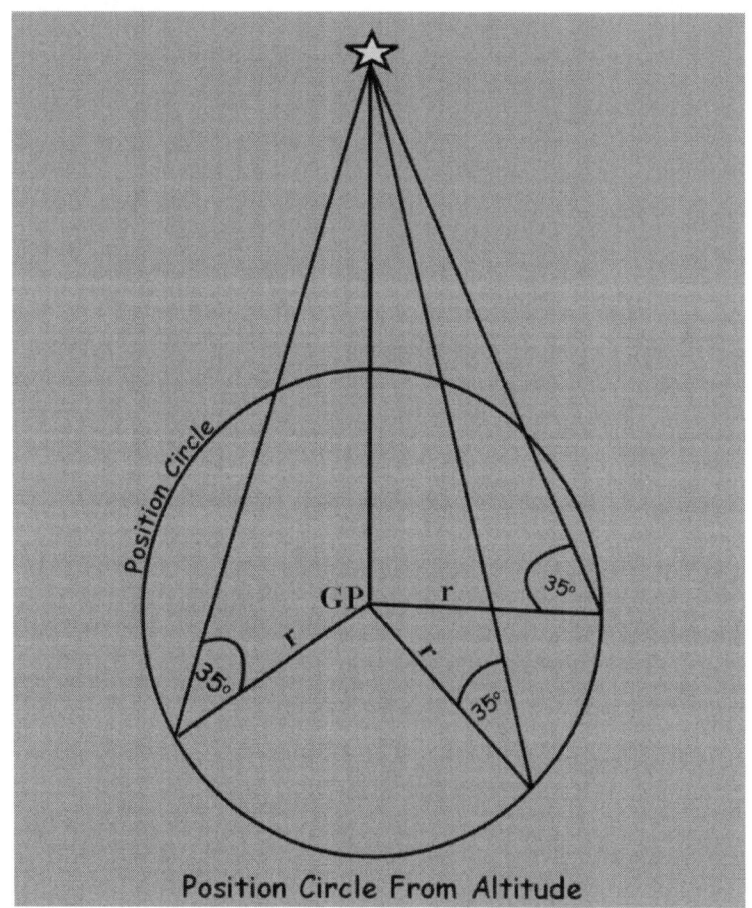

Position Circle From Altitude

At first, it might seem that all we need to do is to observe the azimuth or bearing of the Sun at the same time that we measure its altitude and then draw the line of bearing on the chart along with the position circle. In this way, it would seem that our true position would correspond to the intersection of these lines on the chart. However, there is a problem with this idea which makes it impracticable. Because of the great distance of the Sun from the Earth, the radius of the position circle will be very large (approximately 3000 n.m. or so). A chart on which such a large circle could be drawn would require such a small scale that accurate position-fixing would be impossible.

However, our chart should show our dead-reckoning position (DR) or estimated position (EP) which should be accurate to within a degree of latitude and longitude and this may give us another way of tackling the problem.

Our first step is to establish an 'assumed position' which will be in the close vicinity of the DR or EP position but with the data rounded up to give whole numbers of degrees so that our interpolations will be made easier. We can do this because the DR or EP will be approximate positions and so we can choose another approximate position in the vicinity without too much loss of accuracy.

If we could calculate what the altitude and azimuth would have been at the assumed position at the time that they were accurately measured at the true position, we would then be able to compare the two sets of data and calculate the differences between them in order to establish a position line. (A more thorough treatment of this topic can be found in the book Astro Navigation Demystified).

To determine the altitude and azimuth of a celestial body, we could make calculations by using mathematical formulae; we could compute them with the aid of sight reduction tables; and we could use computer software. However, all of these procedures only tell us what the altitude and azimuth of the chosen body would be at our assumed position which, as previously discussed, is only an approximate position. The only way that we can determine the altitude and azimuth of a celestial body at our true position is by making accurate measurements with a sextant and an azimuth-compass or similar equipment. However, before we can do this, we must obviously be able to locate the body in the sky.

Unlike the Sun and the Moon which are easily identified, the approximate positions of stars and planets need to be established before we can use them for position fixing.

**Sun and Moon.**
The times of rising and setting of the Sun and Moon can be found in the daily pages of the nautical almanac so we know when they will be visible above the horizon and of course, we will have no difficulty in locating them in the sky.

**Stars and Planets.**
**Navigational Stars**. Of all the stars that are visible in the sky, only 59 are considered to be bright enough to be used for navigation; these are known as the Navigational Stars.
A list of navigational stars can be found on page 169 of this book.

**Navigational planets.** Of the eight planets in our solar system, only Venus, Mars, Jupiter and Saturn are considered to be suitable for navigation. The four navigational planets are listed in the daily pages of the Nautical Almanac by their GHA and Declnation.

The times of rising and setting of the Sun and the Moon are listed in the Nautical Almanac but not for the stars and planets and so the navigator has to calculate these when required. In fact, knowledge of the precise times of rising and setting of is not always essential; all that we need to know, in most cases, is whether a selected body will be above our horizon at the time we plan to take a fix and if so, what its approximate position in the sky will be. The 'Where To Look' method which is explained below enables us to calculate this.

## The 'Where To Look' Method.
This is a method of the author's own devising, which enables us to quickly and easily calculate which stars and planets will be above our horizon at the time that we wish to make a 'fix'.

It is quite a useful method especially in the cramped confines of a chart-table, in a rough sea. It is emphasised that this is not intended to be a method of accurately pin-pointing the position of a star or planet but a simple and quick method of establishing whether or not the body is likely to be visible at the time an observation is required and if so, what its approximate position will be in the sky.

**Method.** To find if a star or planet will be above the horizon at our position at the time of the planned observations, we need to take two things into account, its local hour angle and its declination (these terms are fully explained in chapter 2).

**Local Hour Angle (LHA).** For a star or planet to be visible, its meridian must be within 90° east or west of our estimated longitude at the time of the planned observations.

- If a body's LHA is greater than 0° and less than 90° or if LHA is greater than 360° and LHA - 360° is less than 90° then body will be above the western celestial horizon.
- If a body's LHA is greater than 270° but less than 360° then 360° - LHA will be less than 90° and the body will be above the eastern horizon.

We can formulate the above statements as follows:

Body is above western horizon if:
LHA = 0° TO 90° or LHA - 360° = 0° TO 90°
Body is above eastern horizon if:
LHA = 270° to 360° or 360° - LHA = 0° TO 90°

The following diagram demonstrates what has been discussed above.

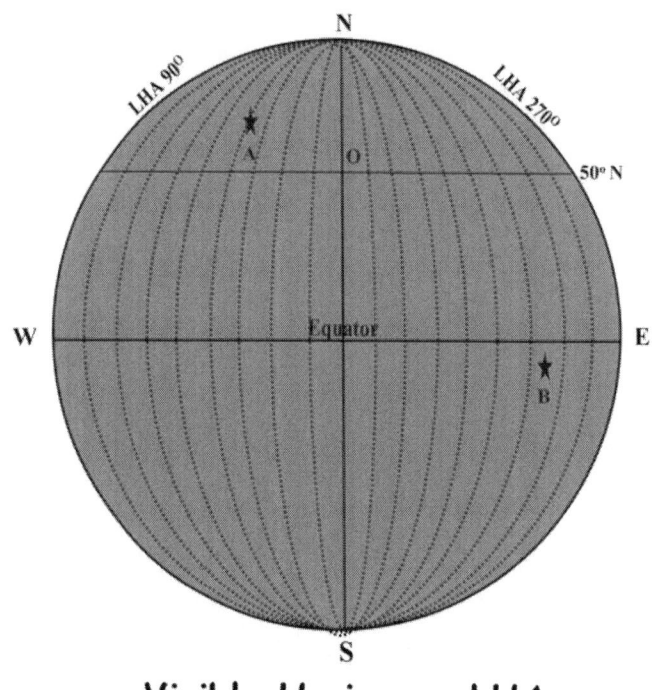

## Visible Horizon - LHA

Point O is the position of an observer on the Earth's surface at latitude 50°N.
NS is the meridian of the observer and in terms of LHA is 0°.
WE is the celestial equator.
The limits of the observer's western and eastern horizons are at LHA 90° and LHA 270° which are both 90° from the observer's meridian.

Suppose that point A in the diagram represents the position of a celestial body whose LHA is 40° and that point B represents the position of another celestial body whose LHA is 295°. Since the LHAs of these bodies is either less than 90° or greater 270°, they will be visible above the horizon to the west and the east respectively.
Bodies whose LHAs are greater than 90° but less than 270° would be below the celestial horizon and therefore would not be visible.

### Calculating LHA

From the above, we can see that our first step in determining the position of a celestial body is to calculate its LHA.

## LHA of A Star.

The LHA of stars is not listed in the Nautical Almanac so we must calculate these ourselves.

The following method can be used to calculate the LHA of a star.

1. From the nautical almanac daily pages, find the Greenwich Hour Angle (GHA) of Aries (to the nearest degree) at the planned time of the observation.
2. From the 'Index to Selected Stars' in the Nautical Almanac, find the Sidereal Hour Angle (SHA) of the star to the nearest degree. (A list of navigational stars can be found on page 169).
3. Calculate your estimated longitude (to the nearest degree) at the planned time of the observation.
4. Combine the SHA, GHA Aries and the estimated longitude to find the approximate LHA.

**Examples.** We will use three celestial bodies, stars X and Y and planet V to demonstrate the method outlined above:

Assumed Position of observer:  Lat. 50°N   Long. 135°W

### Details of Star X:
SHA of X = 182
Declination of X = 55°N
GHA Aries = 335
Longitude of observer = 135°W

### To Calculate LHA of Star X

| | |
|---|---|
| SHA X | 182 |
| GHA Aries | 335 |
| | 517 |
| Long | -135 (subtract if long is west, add if long is east) |
| | 382 |
| | -360 (subtract 360 if LHA is greater than 360) |
| LHA X = | 22° |

Since the LHA of star X is between 0° and 90° it will be visible above the horizon to the West.

∴ LHA = 22°W

**Details of Star Y:**
SHA of Y = 080
Declination of Y = 20°N
GHA Aries = 335
Longitude of observer = 135°W

**To Calculate LHA of Star Y**
SHA Y          080
GHA Aries      335
               415
Long           -135  (subtract if long is west, add if long is east)
LHA Y =        280°

Since the LHA of Y is between 270° and 360° it will be visible above the horizon to the East.

However, from a practical point of view, when the LHA is from 270° to 360°, we need to know how far this is to the east of the observer's meridian. We do this by subtracting the LHA from 360° which in this case is 360° - 280° = 80°E.

∴ LHA = 80°E

**LHA Of Planet V.**
Unlike the stars, the positions of the planets in the celestial sphere are not fixed and for this reason the four navigational planets are listed in the daily pages of the Nautical Almanac by their GHA and declination instead of by their SHA. As discussed in chapter 4, the Navigational planets are: Mars, Saturn, Venus and Jupiter.

The method used to calculate the LHA of a planet is the same as that for a star except that because the GHA is given in the Nautical Almanac, we do not need to calculate it. The following calculations for planet V demonstrate the method.

**Details of Planet V:**
GHA of V = 354
Declination of V = 20°S
Longitude of observer = 135°W

**To Calculate LHA of Planet V:**
GHA           325
Long          -135
**LHA V =     190**

Because the LHA of planet V is greater than 90° but less than 270° it will be below the horizon and therefore it will not be possible to use it for position fixing at the observer's position.

**Notes.**
1. For a position fix, the altitude and azimuth of the celestial bodies being used would be measured within a few seconds of each other so we can assume that the observer's position would be the same for each set of the calculations.
2. For the same reason, the GHA of Aries would have changed by only a minute fraction of a degree.

**To Calculate Azimuth.** Having calculated the LHA of stars X and Y, we can now make a rough approximation of their azimuth by plotting their LHA and declination in relation to the observer's assumed position on the azimuth diagram that we used above.

In the drawing below, we have plotted the positions of X and Y in terms of their LHA and declination which are as follows:

Star X: LHA 22°W, Dec. 65°N
Star Y: LHA 80°E, Dec. 20°N.

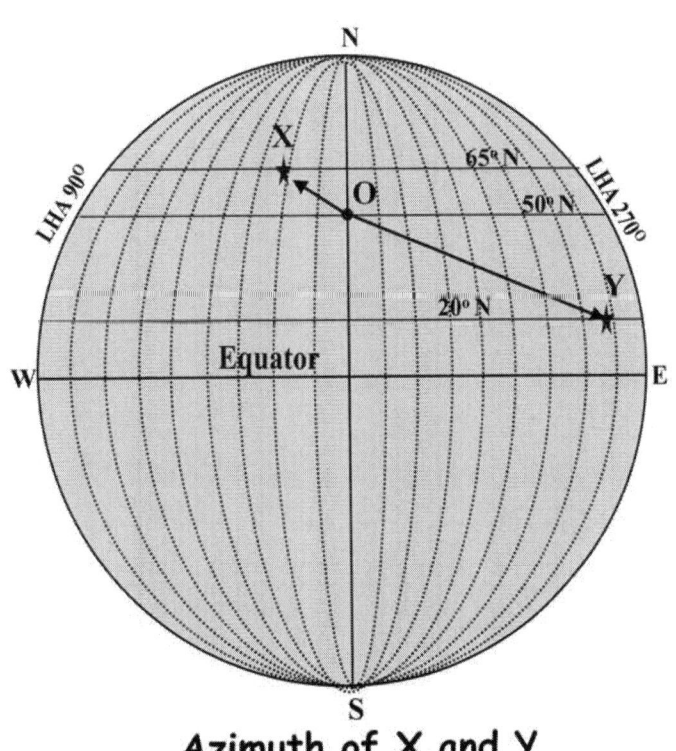

**Azimuth of X and Y**

If we join the positions of X and Y to the position of the observer, we will see that the approximate azimuths are as follows:

Star X: Approximate Azimuth = N30°W

Star Y: Approximate Azimuth = N110°E

Of course we cannot find an accurate solution to a spherical problem with a straight-line, two dimensional, drawing such as this which does not take account of phenomena such as refraction and parallax or the fact that the Earth is not a perfect sphere. However, at this stage we only need an approximate answer to the question "in what direction do we look"?

It is stressed that the diagram above is for illustrative purposes only, it is not drawn to scale and angles are not accurate.

With a little practice, it will become an easy task to estimate the azimuth of a body without the aid of a diagram.

## Declination.

As well as the LHA, we have to take into account the declination of a celestial body. For a celestial body to be visible above the horizon, its declination must be within 90° of the latitude of our position. If our latitude is north, then the declination must be within the range 90° north to (90° - latitude) south. If the latitude is south, then the declination must be within the range 90° south to (90° - latitude) north.

We can formulate the above statements as follows:

Latitude North: visible range = 90°N to (90° - Lat)S.
Latitude South: visible range = 90°S to (90° - Lat)N.

We can explain the reason for this rule with the aid of another diagram which shows the declinations of the bodies X, Y and V from the previous example. Please note that the diagram is for illustrative purposes only; for this reason, it is not drawn to scale and angles are not drawn accurately.

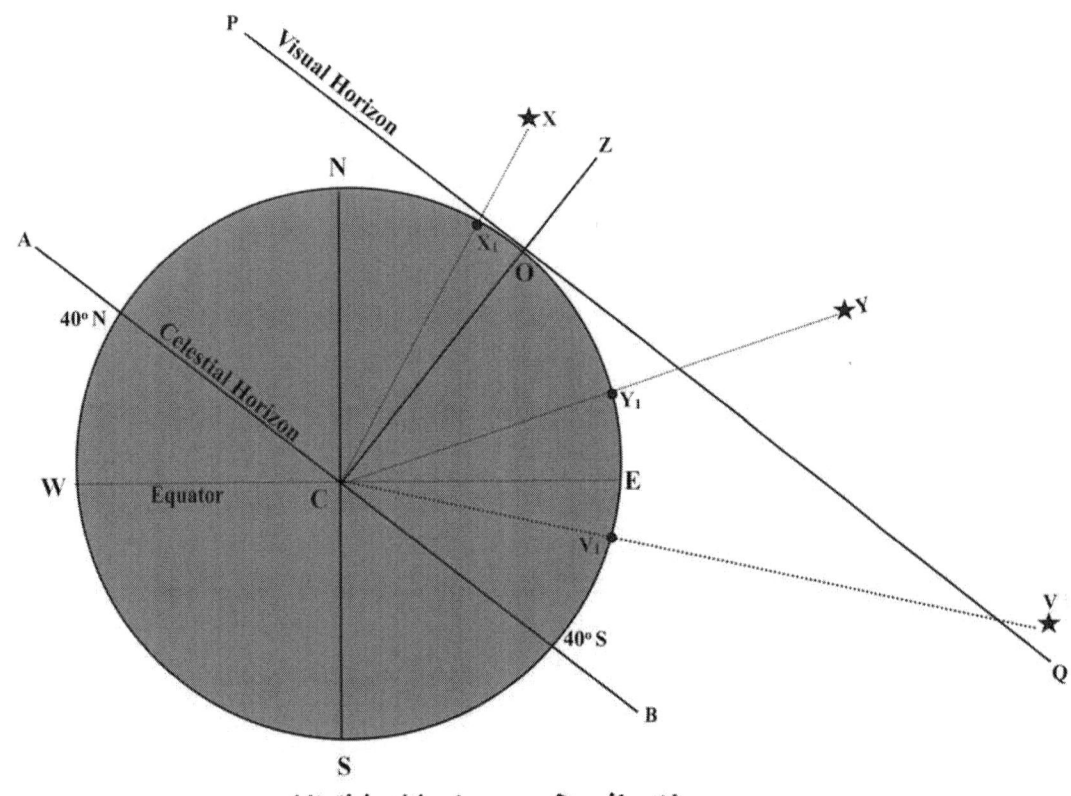

Visible Horizon - Declination

**Summary of data relating to stars X and Y and planet V which are represented in the diagram:**

Assumed Position of observer: Lat. 50°N   Long. 135°W
Star X:   Declination = 65°N   LHA = 22°
Star Y:   Declination = 20°N   LHA = 280°
Planet V: Declination = 10°S   LHA = 190°

In the diagram,
O is the position of the observer at latitude 50° North and Z is the zenith at that position.
The line PQ forms part of the visual horizon at point O and is tangential to the circumference of the Earth.
C is the centre of the Earth and the line CZ is perpendicular the visual horizon.
The line ACB is the celestial horizon which is perpendicular to CZ and therefore horizontal to the visual horizon.

The celestial horizon will cut the circumference of the Earth at latitudes that are 90° to the north and to the south of the observer's latitude (in this example, latitude 40°N and latitude 40°S).

X is the position of the star X in the celestial sphere and $X_1$ is its geographical position on the Earth's surface.

Y is the position of the star Y in the celestial sphere and $Y_1$ is its geographical position.

V is the position of planet V in the celestial sphere and $V_1$ is its geographical position.

The diagram shows that although the geographical positions of X and Y are below the visible horizon and therefore out of sight to the observer, the bodies themselves are above the celestial horizon and therefore since they are within the limits for LHA, they will be visible above the horizon at the observer's assumed position.

The diagram also shows that V would have been above the celestial horizon had it been within limits for LHA.

## Estimating Approximate Altitude.

**Note.** This method enables us to calculate the approximate altitude of a celestial body when it lies over the meridian of the observer; in other words when LHA equals 0° or 360°. When the LHA of a body is greater than 0/360 (i.e. when it is not over the observer's meridian) adjustments have to be made to the calculated altitude as will be shown later.

### Altitude of Stars X and Y and planet V.

We will continue with the example of stars X and Y and planet V for this demonstration. Planet V was found to be outside the visible range for LHA and therefore, below the horizon. However, purely for demonstration purposes, we will continue to estimate what its altitude would have been if it had been within LHA limits.

### Data previously determined:

Latitude of observer's assumed position = 50°N
Star X:   Declination = 65°N   LHA = 22°
Star Y:   Declination = 20°N   LHA = 280°
Planet V: Declination = 10°S   LHA = 190°

We can use the previous diagram to estimate the altitudes of X, Y and V as shown on the next page.

In the diagram, Angle ACN is the angle between the Celestial Horizon and the direction of North and NCX is the angle between the direction of North and the direction of X. Angle ACX is equal to the sum of these angles and represents the altitude of X which is 85°.

Angle BCE is the angle between the Celestial Horizon and the Celestial Equator and ECY is the angle between the direction of East and the direction of Y. BCY is equal to the sum of these angles and represents the altitude of Y which is 60°.

Angle BCV is equal to angle BCE minus the declination of V and represents the altitude of planet V which is 30°. However, in reality, V would not be visible because it is outside the limits for LHA.

**Summary.**
**Altitude X = 85°**
**Altitude Y = 60°**
**Altitude V = 30°**

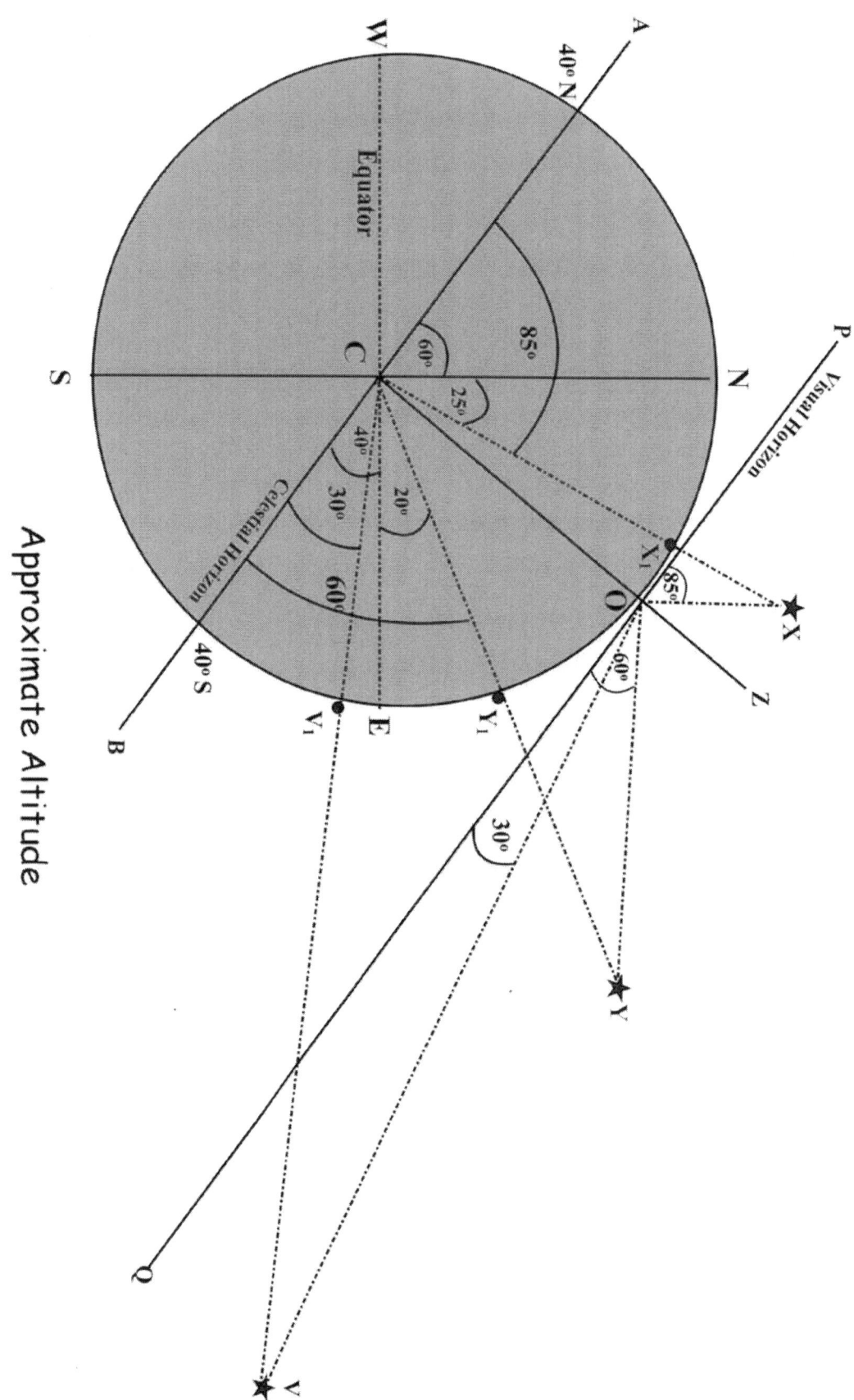

The altitude of a celestial body is 0° when it rises at LHA 90°E and increases to its maximum when LHA is 360°/ 0° before descending again to 0° when it reaches LHA 90°W.

We must remember that the altitudes we have calculated above would apply when the celestial bodies are over the observer's meridian (i.e. when LHA is 0°). Therefore, when the LHA is other than 0°, we have to re-calculate the altitude.

**Re-calculation of altitude of star X for LHA 22°:**
The diagram shows that the altitude of star X is 85°.
∴ Altitude = 85° when LHA = 0°
∴ 90° of LHA = 85° of altitude
But LHA = 22°
∴ 22° of LHA = (85 ÷ 90 x 22) = 20.7° of altitude.
∴ When LHA X = 22°, altitude = 85°- 20.7° = 64.3°
∴ **Estimated altitude of X when LHA is 22° = 64.3°.**

**Re-calculation of Altitude of star Y for LHA 80°:**
The diagram shows that the altitude of star Y is 60°.
∴ Altitude = 60° when LHA = 0°
∴ 90° of LHA = 60° of altitude
But LHA = 280° = 80° E.
∴ 80° of LHA = (60 ÷ 90 x 80) = 53° of altitude.
∴ When LHA = 80°, altitude = 60° - 53° = 7°
∴ **Estimated altitude of Y when LHA is 80° = 7°.**

We cannot make a recalculation for planet V because its LHA is outside LHA limits.

From the above findings, we can formulate rules for estimating altitude as follows.
**Case 1. When the observer is in the Northern Hemisphere.**
**Rule 1.When the observer and the celestial body are both in the Northern Hemisphere:**
- If the declination of the body is to the south of the latitude of the observer then altitude is measured from the southern part of the celestial horizon and altitude equals the angle between the celestial

horizon and the celestial equator plus the angle between the celestial equator and the declination of the body.
- If the declination of the body is to the north of the latitude of the observer then altitude is measured from the northern part of the celestial horizon and the altitude equals the angle between the celestial horizon and the North Pole plus the angle between the North Pole and (90 - Dec).

**Rule 2. When the observer is in the Northern Hemisphere and the declination of the body is south:**

The altitude is measured from the southern part of the celestial horizon and the altitude equals the angle between the celestial horizon and the celestial equator minus the declination of the body.

**Note.** If the declination of the body is below the celestial horizon then the body will not be visible at the observer's position.

**Case 2. When the observer is in the Southern Hemisphere.** In this case, we can simply take the rules for the Northern Hemisphere and 'turn them upside-down,' so to speak.

**Rule 1. When the observer and the celestial body are both in the Southern Hemisphere:**
- If the declination of the body is to the north of the latitude of the observer then altitude is measured from the northern part of the celestial horizon and altitude equals the angle between the celestial horizon and the celestial equator plus the angle between the celestial equator and the declination of the body.
- If the declination of the body is to the south of the latitude of the observer then altitude is measured from the southern part of the celestial horizon and altitude equals the angle between the celestial horizon and the south pole plus the angle between the south pole and (90 - Dec).

**Rule 2. When the observer is in the Southern Hemisphere and the declination of the body is north:**

The altitude is measured from the northern part of the celestial horizon and the altitude equals the angle between the celestial horizon and the celestial equator minus the declination of the body.

**Note.** If the declination of the body is below the celestial horizon then the body will not be visible at the observer's position.

Once again, we must remember that the 'Where to Look' method gives straight line solutions to a spherical problem and does not take account of phenomena such as refraction and parallax so it can only give us approximate answers to the problems of azimuth and altitude. However, it does give us a good idea where to look in the sky for a celestial body that we have selected to use for an accurate position fix.

**Accuracy of the method.** Notwithstanding the limitations discussed above, it has been estimated through practice that the 'Where to Look' Method is accurate to within $\pm 10°$ which is adequate for its intended task.

Now that the explanations and demonstrations have been completed, we can summarize the complete method by estimating the azimuth and altitude of the star Alioth and the planet Jupiter in a set scenario.

**Full Example.** To estimate the approximate azimuth and altitude of the star Alioth and the planet Jupiter.
**Scenario.**
A navigator intends to use the star Alioth and the planet Jupiter for a position fix. Before he can do this, he needs to locate these bodies in the sky. This he does this by using the 'Where to Look' method to calculate their approximate azimuth and altitude as demonstrated below:
He bases his calculations on the following data:
Assumed Position of observer: Lat. 30°N   Long. 45°W
SHA of Alioth = 166.  Declination = 56°N
GHA Aries = 250
GHA of Jupiter = 026.  Declination = 17°N

**Step 1. Calculate LHA.**
**Alioth.**

| | |
|---|---|
| SHA Alioth | 166 |
| GHA Aries | 250 |
| | 416 |
| Long | -45 (subtract if long is west, add if long is east) |
| | 371 (subtract 360 if LHA is greater than 360) |
| | -360 |
| LHA Alioth | 11 = 11°(W) |

**Jupiter.**
**Calculate LHA**

GHA  026
Long  -45
LHA =  -19 = (19°E) (Explanation. Longitude is 45° to the west of the Greenwich Meridian and the Greenwich Hour Angle is 26° to the west of Greenwich which is 19° to the east of the observer's meridian; so the Local Hour Angle must be 19°E).

So we can make the following rule: +LHA = West and -LHA = East.

**Step 2. Calculate whether or not the bodies are above the horizon.**
The next step is to determine whether or not the bodies are above the horizon. The rules for this are repeated below:

**Reminder:**
The body is above western horizon if:
LHA = 0° TO 90° or LHA - 360° = 0° TO 90°
The body is above eastern horizon if:
LHA = 270° to 360° or 360° - LHA = 0° TO 90°

**Conclusion.** The LHA of Alioth is 11°W and so we can conclude that it will be above the horizon to the west. The LHA of Jupiter is 19°E so we can conclude that it is above the horizon to the east.

**Step 3. Estimate Azimuth.**
**To Estimate Azimuth.** Having calculated the LHA of Alioth and Jupiter, we can now make a rough approximation of their azimuths by plotting their LHA and declination in relation to the observer's assumed position using the azimuth diagram that we used earlier.

In the drawing, we have plotted the positions of Alioth and Jupiter in terms of their LHA and declination which are as follows:
Alioth: LHA 11°W, Dec. 56°N
Jupiter: LHA 19°E, Dec. 17°N.

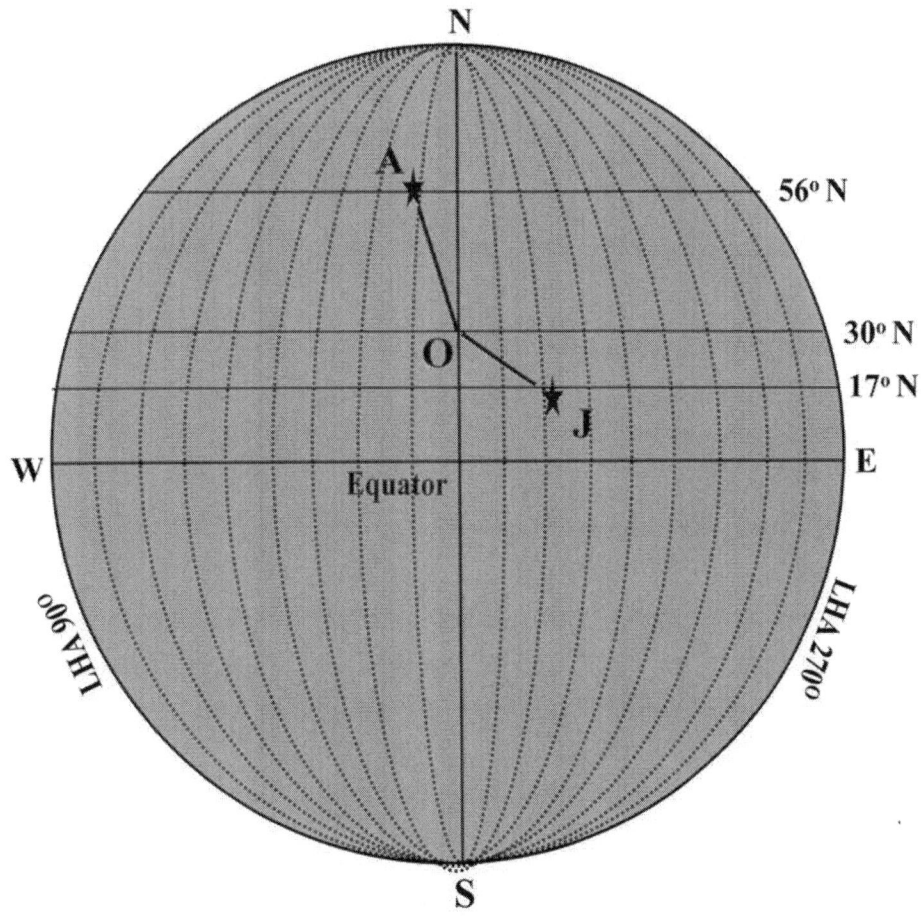

**Alioth and Jupiter - Azimuth**

If we draw lines joining the positions of Alioth and Jupiter to the position of the observer, we will see that the approximate azimuths are as follows:
Alioth: Azimuth = N15°W
Jupiter : Azimuth = N110°E

Of course, as discussed earlier, we cannot find an accurate solution to a spherical problem with a straight-line, two dimensional drawing such as this which does not take account of the fact that the Earth is not a perfect sphere and does not allow for phenomena such as refraction and parallax . However, at this stage we only need an approximate answer to the question "in what direction do we look"?
It is also stressed that the drawing above is for illustrative purposes only and therefore, it is not drawn to scale and angles are not accurately drawn.

With a little practice, it will become an easy task to estimate the azimuth of a body without the aid of a diagram.

**Step 4. Are the bodies above the horizon to the north and south?**
The declinations of Alioth and Jupiter are as follows:
Alioth: Dec. 56°N
Jupiter: Dec. 17°N.
**Reminder:**
  Latitude North: visible range = 90°N to (90° − Lat)S.
  Latitude South: visible range = 90°S to (90° − Lat)N.
Since the LHAs and declinations of Alioth and Jupiter satisfy the above rules, we can conclude that they will be visible above the horizon.

**Step 5. Calculate Approximate Altitude.**
**Note.** This method enables us to calculate the approximate altitude of a celestial body when it lies over the meridian of the observer; in other words when LHA equals 0° or 360°. When the body is not on the observer's meridian, adjustments have to be made to the calculated altitude.
**Data previously determined:**
Latitude of observer = 30°N
Declination of Alioth = 56°N, LHA = 11°W
Declination of Jupiter = 17°N, LHA = 19°E
We can make a rough approximation of the altitudes of Alioth and Jupiter by plotting their declinations in relation to the observer's approximate latitude on the altitude diagram that we used earlier.

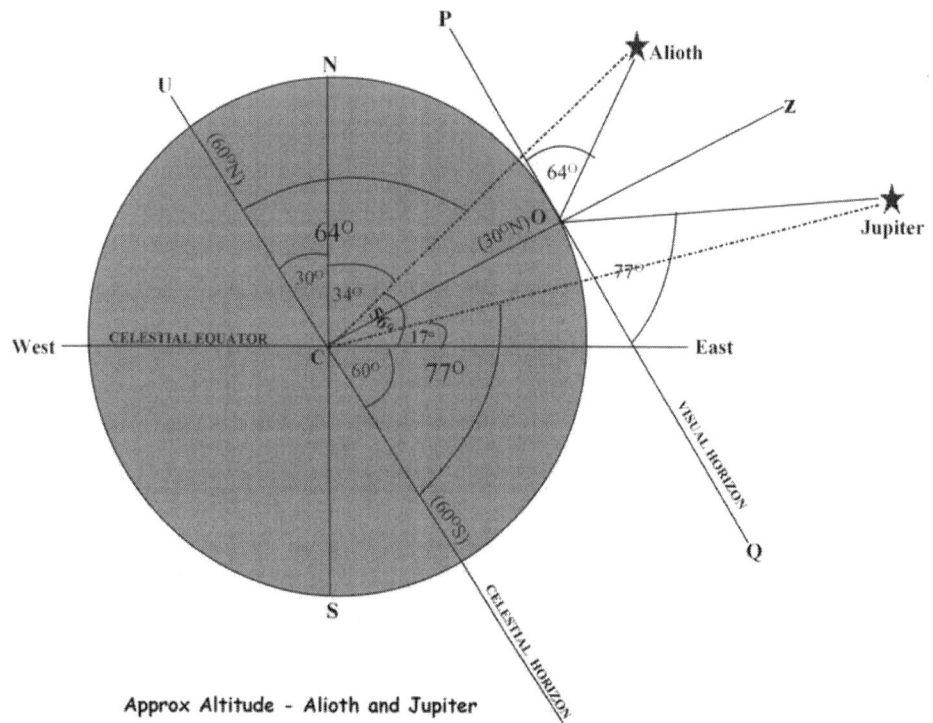

**Approx Altitude - Alioth and Jupiter**

### Approximate Altitude of Jupiter.
The diagram shows that the altitude of Jupiter is 77°.
However, we must remember that this is the altitude when Jupiter is on the observer's meridian. (The altitude of a celestial body is 0° when it rises at LHA 90°E and increases to its maximum when LHA is 360°/ 0°).

### Calculations:
Altitude = 77° when LHA = 0°
∴ 90° of LHA = 77° of altitude
∴ 19° of LHA = (77 ÷ 90 x 19) = 16.2° of altitude.
∴ When LHA Jupiter = 19°E, altitude = 77°- 16.2° = 60.8°
∴ Approx. altitude of Jupiter = 60.8°.

### Approximate Altitude of Alioth.
The diagram also shows that, in this example, the altitude of Alioth is 64°.
So by the same reasoning as above:
90° of LHA = 64° of altitude
∴ 11° of LHA = (64 ÷ 90 x 11) = 7.8° of altitude.
∴ When LHA Alioth = 11°W, altitude = 64°- 7.8° = 56.2°
∴ Approx. altitude of Alioth = 56.2°.

### Summary of approximate positions of Alioth and Jupiter.
Alioth: Azimuth = N15°W   Altitude = 56.2°.
Jupiter: Azimuth = N110°E   Altitude = 60.8°.

### Summary of Data Gathered.
### Alioth.
Assumed Position of Observer: Lat. 30°N   Long. 45°W
SHA of Alioth = 166
GHA Aries = 250
LHA of Alioth: 11°W
Declination of Alioth = 56°N
Approximate Azimuth: N15°W
Approximate Altitude: 56.2°.

### Jupiter.
Assumed Position of Observer: Lat. 30°N   Long. 45°W
GHA of Jupiter: 026

LHA of Jupiter:
Declination of Jupiter: 17°N
Approximate Azimuth: N110°E
Approximate Altitude: 60.8°

**Using The Values Of Altitude And Azimuth To Calculate Our Position.**
Our aim in, astro navigation, is to calculate the azimuth and altitude of a celestial body from our assumed position and then to compare the results with the azimuth and altitude accurately measured at our true position. With this information we will then be able to establish a position line from the assumed position to the true position. Such a position line is called 'the intercept'.

Once we have located the bodies by the 'where to look' method, we are in a position to measure them accurately from our true position using a sextant and azimuth compass.

The next step is to calculate what the azimuth and altitude would have been at the assumed position at the precise time that they were measured at the true position.

As previously discussed, there are several ways of calculating the azimuth and altitude at the assumed position such as by the use of sight reduction methods and software solutions. However, the traditional method is by the use of spherical trigonometry which is demonstrated below.

**Accurately Calculating Azimuth and Altitude at the Assumed Position by Spherical Trigonometry.**
(Note. An exposition of spherical trigonometry is given in chapter 9).

Please consider the following diagram.

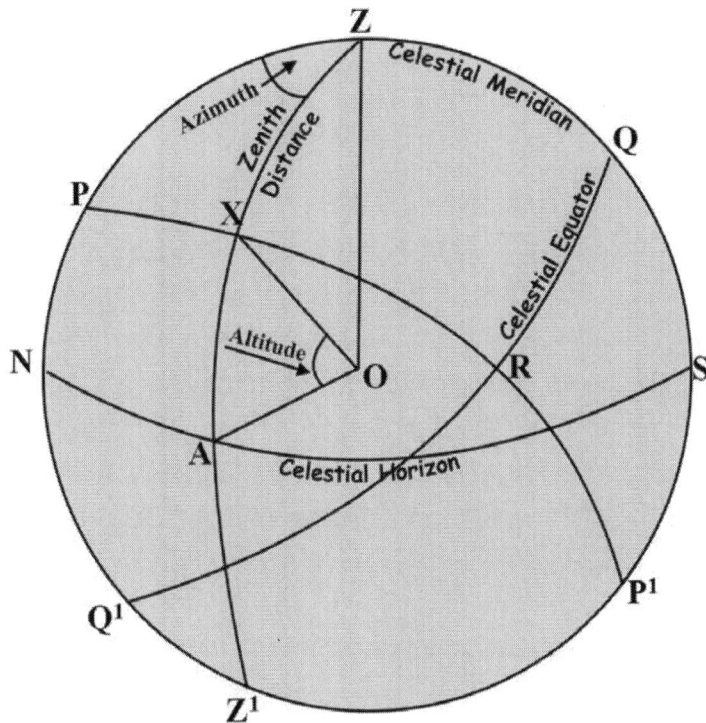

**Altitude, Azimuth and Zenith Distance**

QZ is the angular distance from the zenith to the Celestial Equator and is equal to the latitude of the observer.

PZ is the angular distance from the Celestial North Pole to the zenith of the observer and is equal to 90° - Lat.

XR is the angular distance from the celestial body to the Celestial Equator and is equal to the declination of the body.

PX is the angular distance from the Celestial North Pole to the celestial body and is equal to 90° - Dec.

The angle AOX is the altitude of the Celestial Body.

The angle XOZ is the complement of AOX so that AOX plus XOZ = 90°.

ZX is the Zenith Distance and is equal to 90° - AX.

Therefore, 90° - ZX - is equal to the altitude of the body.

The angle ZPX is equal to the Local Hour Angle of the Celestial Body with respect to the observer's meridian.

The angle PZX is the azimuth of the body with respect to the observer's meridian.

**Summary.**
PX = 90° - Dec.
PZ = 90° - Lat.
ZX = 90° - Alt.
Alt = 90° - ZX
‹PZX = Azimuth.

Since our aim is to calculate the azimuth and altitude of a celestial body we must solve the spherical triangle PZX. Specifically, we must calculate the angular distance of side ZX so that we can find the altitude and we must calculate the angle PZX so that we can find the azimuth.
However, Because PZX is on the surface of an imaginary sphere, we cannot solve this triangle by the use of 'straight line trigonometry'; instead we must resort to the use of 'spherical trigonometry' which is explained in chapter 9. The following examples show how the calculations can be made by the use of spherical trigonometry.

**Examples of the use of spherical trigonometry to calculate the azimuth and altitude of celestial bodies.** In these examples, we use the data gathered from previous examples. Our task is to calculate the altitude and azimuth of Alioth and Jupiter at the assumed position.

**Example 1.   Alioth**
Use spherical trigonometry to calculate the azimuth and altitude of Alioth at the assumed position.
Data given earlier:
Assumed Position:  Lat. 30°N   Long. 45°W
Position of Alioth:  SHA: 166  Declination: 56°N
GHA Aries:  250

**Step 1.   Calculate LHA**

| | |
|---|---|
| SHA Alioth | 166 |
| GHA Aries | 250 |
| | 416 |
| Long | -45 |
| | 371 |
| | -360 |
| LHA = | 11 = 11°W |

**Step 2. Calculate PZ/PX/ZPX**

PZ = 90° - 30° = 60°    ∴ PZ = 60°
PX = 90° - 56° = 34°    ∴ PX = 34°
ZPX = LHA = 11° (W)

**Step 3.   Calculate Zenith Distance (ZX).**

As explained in chapter 9, the formula for calculating side ZX is:
Cos (ZX) = [Cos(PZ) . Cos(PX)] + [Sin(PZ) . Sin(PX) . Cos(ZPX)]


∴ To calculate zenith distance of Alioth:
Cos (ZX) = [Cos(PZ) . Cos(PX)] + [Sin(PZ) . Sin(PX) . Cos(ZPX)]
         = [Cos(60°) . Cos(34°)] + [Sin(60°) . Sin(34°) . Cos(11°)]
         = [0.5 x 0.829} + [0.866 x 0.559 x 0.982]
         = 0.415 + 0.475
Cos (ZX) = 0.89
∴ ZX    = Cos$^{-1}$ (0.89)
        = 27°

**Step 4.   Calculate Altitude.**

Altitude = 90° - ZX
         = 90° - 27°
         = 63°

**Step 5.   Calculate Azimuth (PZX)**

As explained in chapter 9, the formula for calculating angle PZX is:

Cos PZX = $\frac{Cos(PX) - [Cos(ZX) . Cos(PZ)]}{[Sin(ZX) . Sin(PZ)]}$

∴ To calculate azimuth of Alioth:

Cos PZX = $\frac{Cos(PX) - [Cos(ZX) . Cos(PZ)]}{[Sin(ZX) . Sin(PZ)]}$

= $\frac{Cos(34) - [Cos(27) . Cos(60]}{[Sin(27) . Sin(60)]}$

= $\frac{0.829 - [ 0.89 \times 0.5]}{0.454 \times 0.866}$

= $\frac{0.829 - 0.445}{0.393}$

= $\frac{0.384}{0.393}$

Cos(PZX) = 0.977

∴ PZX = Cos⁻¹(0.977) = 12.31

∴ Azimuth = N12°W (since LHA is west)

In terms of bearing, the azimuth is 348°.

**Step 6. Summarize position of Alioth**

LHA = 11° (W)

Declination = 56°N

Azimuth at assumed position = N12°W

Altitude at assumed position = 63°

**Example 2.** Use spherical trigonometry to calculate the azimuth and altitude of Jupiter at the assumed position.

**Data established earlier:**

Assumed Position of Observer: Lat. 30°N   Long. 45°W

GHA of Jupiter: 026

Declination of Jupiter: 17°N

**Step 1.   Calculate LHA**

GHA           026

Long          -45

LHA =        -19   = (19E) [LHA<180° = West,  LHA>180° = East]

**Note.**   Because the GHA for planets is given in the Nautical Almanac, we do not need to calculate it.

**Step 2. Calculate PZ/PX/ZPX**

PX = 90° - Dec.

PZ = 90° - Lat.

ZX = 90° - Alt.

Alt = 90° - ZX

<PZX = Azimuth.

Calculate PZ:  PZ = 90° - Lat.
              = 90° - 30° = 60°

Calculate PX:  PX = 90° - Dec
              = 90° - 17° = 73°

ZPX = LHA = 19°(E)

**Step 3. Calculate Zenith Distance (ZX).**

$$\cos(ZX) = [\cos(PZ) \cdot \cos(PX)] + [\sin(PZ) \cdot \sin(PX) \cdot \cos(ZPX)]$$
$$= [\cos(60°) \cdot \cos(73°)] + [\sin(60°) \cdot \sin(73°) \cdot \cos(19°)]$$
$$= [0.5 \times 0.292] + [0.866 \times 0.956 \times 0.945]$$
$$= 0.146 + 0.475$$

$\cos(ZX) = 0.782$

$ZX = \cos^{-1}(0.782)$
$= 38.5°$

**Step 4. Calculate Altitude.**

Altitude $= 90° - ZX$
$= 90° - 38.5°$
$= 51.5°$

**Step 5. Calculate Azimuth (PZX)**

$$\cos PZX = \frac{\cos(PX) - [\cos(ZX) \cdot \cos(PZ)]}{[\sin(ZX) \cdot \sin(PZ)]}$$

$$= \frac{\cos(73) - [\cos(38.5) \cdot \cos(60)]}{[\sin(38.5) \cdot \sin(60)]}$$

$$= \frac{0.292 - [0.783 \times 0.5]}{0.622 \times 0.866}$$

$$= \frac{0.292 - 0.391}{0.538}$$

$$= \frac{-0.099}{0.538}$$

$\cos(PZX) = -0.184$

$\therefore PZX = \cos^{-1}(0.184) = 100.6 \approx 101°$

$\therefore$ Azimuth = N101°E  (Bearing 101°)

**Summary of data relating to Jupiter**

LHA = 19° (E)
Declination = 17°N
Azimuth = N101°E
Altitude = 51.5°

**Summary of findings.**

1. Approximate altitude and azimuth calculated by the 'Where to Look' method:
Alioth: Azimuth = N15°W   Altitude = 56.2°.
Jupiter: Azimuth = N110°E   Altitude = 60.8°.

2. Altitude and azimuth calculated using spherical trigonometry:
Alioth: Azimuth = N12°W   Altitude = 63°.
Jupiter: Azimuth = N101°E   Altitude = 51.5°.

**Conclusion.** It can be seen that the approximate azimuths and altitudes which we calculated earlier for Alioth and for Jupiter are within the $\pm 10°$ accuracy limit of those that we have now calculated accurately by spherical trigonometry thereby proving the 'where to look' method of making approximations.

In practice, the navigator could use the 'where to look' method to locate his chosen celestial bodies so that he could accurately measure their azimuth and altitude from the true position. He would then calculate what their azimuth and altitude would have been at the assumed position at the precise time that the measurements were taken at the true position. Armed with this data he would then be able to compare the two sets of data to enable him to calculate the differences between them so that he could establish 'position lines' for the celestial bodies by the method explained in detail in the book 'Astro Navigation Demystified'.

**Note.** The SHA, declination and magnitude of each navigational star is listed in the 'Index to Selected Stars' which is contained in the Nautical Almanac. For convenience, a table of navigational stars is to be found on page 169.

# Chapter 9
# Introduction to Spherical Trigonometry.

In practical astro navigation, we mostly rely on the use of tables of computed data and rote-learned procedures. We can operate quite efficiently in this way for most of the time; however, for those who wish to develop a thorough understanding of the subject, it is important to study the principles of spherical trigonometry which underpin astro navigation.

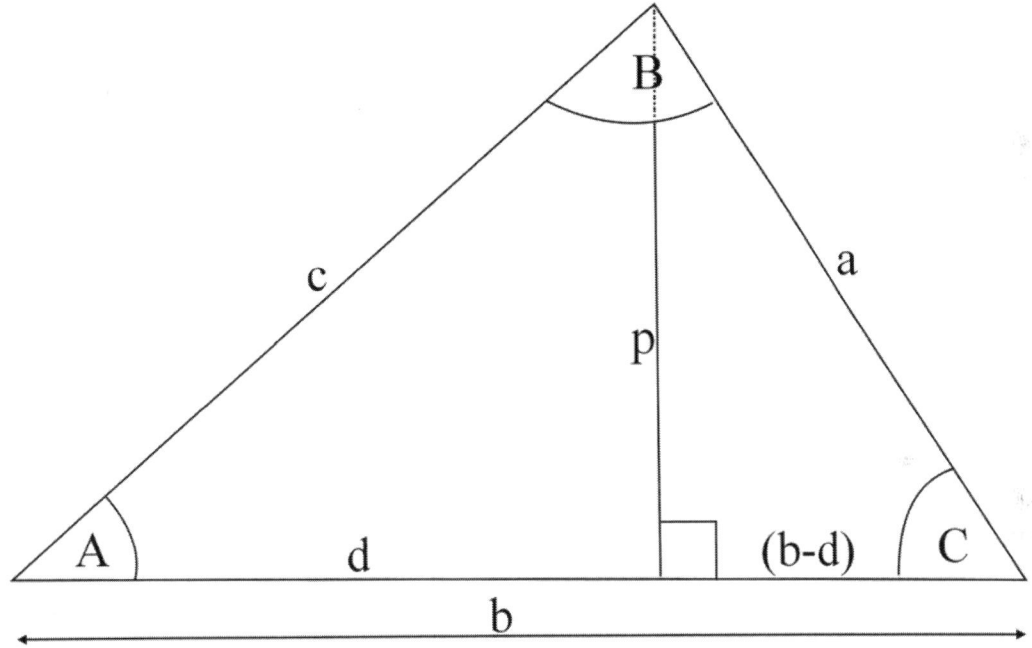

In the diagram above,
$\cos C = d/a$ and $d = a \cos C$
$p^2 = c^2 - (b-d)^2$ and $p^2 = a^2 - d^2$
☐ $a^2 - d^2 = c^2 - (b-d)^2$
☐ $a^2 - d^2 = c^2 - b^2 + 2bd - d^2$
☐ $a^2 = c^2 - b^2 + 2bd$
☐ $a^2 = c^2 - b^2 + 2b.a \cos C$ (since $d = a \cos C$)
☐ $c^2 = a^2 + b^2 - 2b.a \cos C$

This is the **cosine rule** for 'flat' triangles but does this rule also apply to spherical triangles?

The next diagram shows a spherical triangle ABC formed by the intersection of three circles with their common centre O at the centre of the sphere.

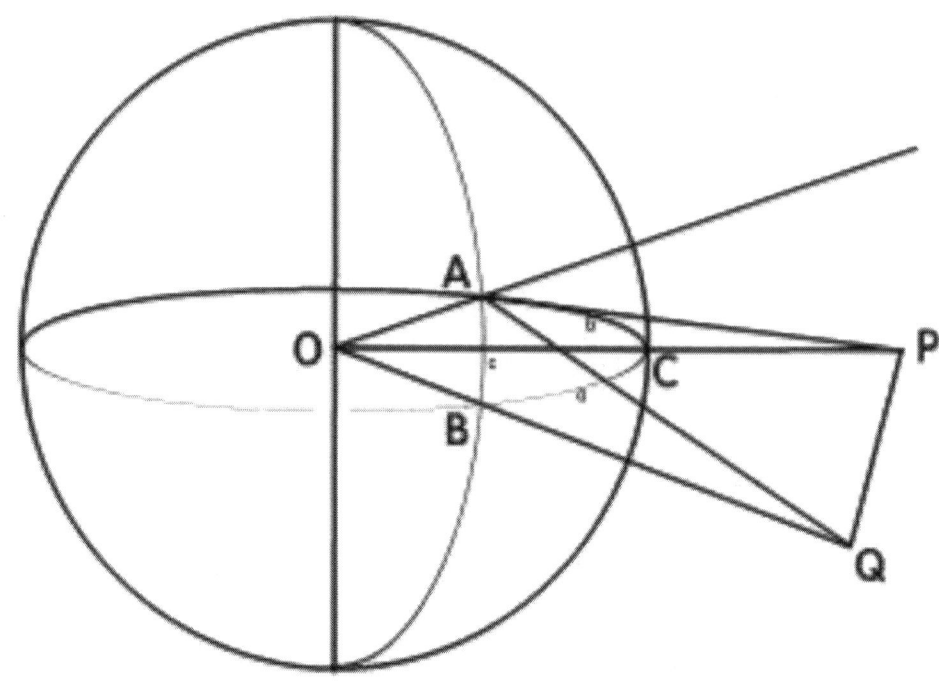

The edges of the 'flat' planes in which the sides a, b, & c lie, meet along OA, OQ and OP.

The edges AQ and AP just graze the great circles of the arcs c and b at A; that is, AQ is the tangent to c and AP is the tangent to b. So angles OAQ and OAP are really right angles although it is impossible to draw them as such in the 'flat'.

The edges of the three planes in which a, b, & c lie form a flat (two-dimensional) triangle PAQ, of which the apical angle PAQ is equivalent to the angle A of the spherical triangle.

Now, by the rule for 'flat' triangles, we have:
$$PQ^2 = PO^2 + QO^2 - 2PO \cdot QO \cdot \cos(a)$$
and $$PQ^2 = PA^2 + QA^2 - 2PA \cdot QA \cdot \cos(A)$$

$\Rightarrow$    $[PO^2 - PA^2] + [QO^2 - QA^2] - [2PO.QO.Cos(a)] + [2PA . QA .Cos(A)] = 0$

We also have:

$PO^2 - PA^2 = AO^2$ and $QO^2 - QA^2 = AO^2$

$\Rightarrow$    $PO^2 - PA^2] + [QO^2 - QA^2] = 2AO^2$

Substituting $2AO^2$ in the equation gives us:

$2PO . QO . Cos(a) = 2AO^2 + 2PA . QA .Cos(A)$

Dividing through by $2PO . QO$, we have:

$$Cos(a) = \frac{AO}{PO} . \frac{AO}{QO} + \frac{PA}{PO} . \frac{QA}{QO} Cos(A)$$

$= [Cos(POA) . Cos(QOA)] + [Sin(POA) . Sin(QOA) . Cos(A)]$

$= [Cos(b) . Cos(c)] + [Sin(b) . Sin(c) . Cos(A)]$

Hence, the formula for finding the third side (a) of a spherical triangle when the other two sides (b and c) are known together with the included angle (A) is: $Cos(a) = [Cos(b) . Cos(c)] + [Sin(b) . Sin(c) .Cos(A)]$

**(This is the cosine rule for spherical triangles).**

We apply this rule when solving the spherical triangle PZX For example, in the next diagram, if we wish to find the side ZX in the spherical triangle PZX, we have:

$Cos ZX = [Cos PZ. Cos PX] + [Sin PZ. Sin PX. Cos ZPX]$

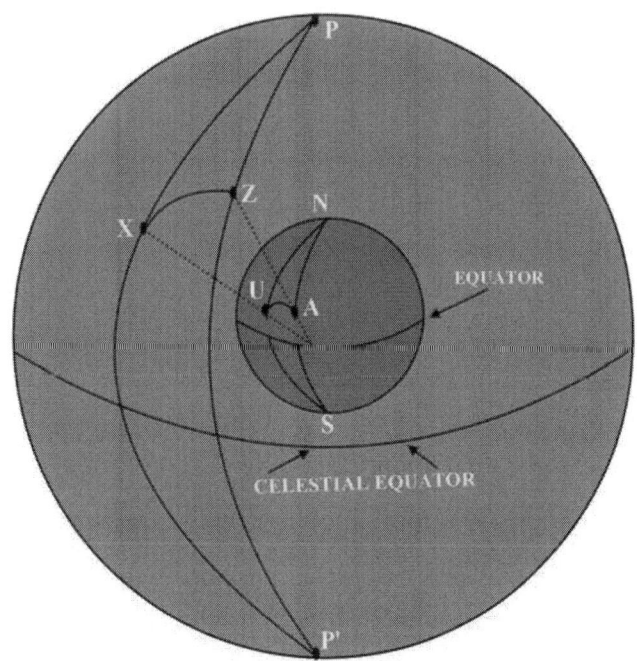

**The PZX Triangle**

**Example.**

Given:
PZ = 60°
PX = 34°
ZPX = 11° (W)
Calculate Zenith Distance (ZX).

$\text{Cos}(ZX) = [\text{Cos}(PZ) \cdot \text{Cos}(PX)] + [\text{Sin}(PZ) \cdot \text{Sin}(PX) \cdot \text{Cos}(ZPX)]$

∴ To calculate zenith distance of Alioth:

$$\begin{aligned}
\text{Cos}(ZX) &= [\text{Cos}(PZ) \cdot \text{Cos}(PX)] + [\text{Sin}(PZ) \cdot \text{Sin}(PX) \cdot \text{Cos}(ZPX)] \\
&= [\text{Cos}(60°) \cdot \text{Cos}(34°)] + [\text{Sin}(60°) \cdot \text{Sin}(34°) \cdot \text{Cos}(11°)] \\
&= [0.5 \times 0.829] + [0.866 \times 0.559 \times 0.982] \\
&= 0.415 + 0.475
\end{aligned}$$

$\text{Cos}(ZX) = 0.89$

∴ $ZX = \text{Cos}^{-1}(0.89)$
$\phantom{\therefore ZX\ } = 27°$

# Navigational Stars

The following is a list of the 59 navigational stars together with their magnitude, sidereal hour angle (SHA) and declination.

Note. Star Magnitude. It may seem illogical but the higher the value of its magnitude, the lower the brightness of the star. For example, the brightest star in the sky is Sirius with a magnitude of -1.47 while the dullest navigational star is Zubenelgenubi with a magnitude of 3.28. By comparison, the magnitude of the Sun is –27, a full moon is –13 and Venus, the brightest planet, is –5.

| Name | Mag. | SHA | Dec.(°) |
|---|---|---|---|
| Acamar | 3.2 | 315 | S.40 |
| Achernar | 0.5 | 335 | S.57 |
| Acrux | 1.3 | 173 | S.63 |
| Adhara | 1.5 | 256 | S.29 |
| Aldebaran | 0.9 (Var) | 291 | N.17 |
| Alioth | 1.8 | 167 | N.56 |
| Alkaid | 1.9 | 153 | N.49 |
| Al Na'ir | 1.7 | 28 | S.47 |
| Alnilam | 1.7 | 276 | S.1 |
| Alphard | 2.0 | 218 | S.09 |
| Alphecca | 2.2 | 126 | N.27 |
| Alpheratz | 2.1 | 358 | N.29 |
| Altair | 0.8 | 62 | N.9 |
| Ankaa | 2.4 | 353 | S.42 |
| Antares | 1.09 | 113 | S.26 |
| Arcturus | 0.0 | 146 | N.19 |
| Atria | 1.9 | 108 | S.69 |
| Avior | 2.4 | 234 | S.59 |
| Bellatrix | 1.6 | 279 | N.6 |
| Betelgeus | 0.58(Var) | 271 | N.7 |
| Canopus | -0.72 | 264 | S.53 |
| Capella | 0.71 | 281 | N.46 |
| Deneb | 1.25 | 050 | N.45 |
| Denebola | 2.14 | 183 | N.15 |
| Diphdar | 2.04 | 349 | S.18 |

| Dubhe | 1.87 | 194 | N.62 |
|---|---|---|---|
| Elnath | 1.68 | 279 | N.29 |
| Eltanin | 2.23 | 091 | N.51 |
| Enif | 2.4 | 034 | N.10 |
| Fomalhaut | 1.16 | 016 | S.30 |
| Gacrux | 1.63 | 172 | S.57 |
| Gienah | 2.8 | 176 | S.17 |
| Hadar | 0.6 | 149 | S.60 |
| Hamal | 2.00 | 328 | N.23 |
| Kaus Australis | 1.8 | 084 | S.34 |
| Kochab | 2.08 | 137 | N.74 |
| Markab | 2.49 | 014 | N.15 |
| Menkar | 2.5 | 315 | N.04 |
| Menkent | 2.06 | 149 | S.36 |
| Miaplacidus | 1.7 | 222 | S.70 |
| Mirfak | 1.82 | 309 | N.50 |
| Nunki | 2.06 | 076 | S.26 |
| Peacock | 1.91 | 054 | S.57 |
| Polaris | 2.01 | 319 | N.89 |
| Pollux | 1.15 | 244 | N.28 |
| Procyon | 0.34 | 245 | N.05 |
| Rasalhague | 2.10 | 091 | N.13 |
| Regulus | 1.35 | 208 | N.12 |
| Rigel | 0.12 | 282 | S.08 |
| Rigil Kentaurus | -0.01 | 146 | N.19 |
| Sabik | 2.43 | 103 | S.16 |
| Schedar | 2.25 | 350 | N.56 |
| Shaula | 1.62 | 097 | S.37 |
| Sirius | -1.47 | 259 | S.17 |
| Spica | 1.04 | 159 | S.11 |
| Sulhail | 2.23 | 223 | S.43 |
| Vega | 0.03 | 081 | N.39 |
| Zubenelgenubi | 3.28 | 138 | S.16 |

Note. This list includes Polaris which is not listed as a navigation star in the nautical almanac issued by the United Kingdom Hydrographic Office.

© Copyright Jack Case 2015

However, it has always played an important role in navigation; not only because it indicates the direction of north but also because it can be used for position fixing particularly in the polar-regions and for this reason, it is listed in the American Practical Navigator as a navigation star

# INDEX

| | |
|---|---|
| Acrux | 110 |
| Aldebaran | 81 |
| Alioth | 51 |
| Alkaid | 51 |
| Alphecca | 101 |
| Alpheratz | 63 |
| Altair | 69 |
| Altitude | 15, 22, 148 |
| Andromeda | 62 |
| Anilam | 87 |
| Antares | 97 |
| Apparent Noon | 129 |
| Apparent Solar Time | 124 |
| Aquarius | 74 |
| Aries | 92 |
| Arturus | 101 |
| Assumed Position | 139 |
| Astro Navigation | 5, 8, 9, 138 |
| Astronomical Day | 129 |
| Auriga | 85 |
| Autumnal Equinox | 13 |
| Axial Precession | 12 |
| Axial Tilt | 12 |
| Azimuth | 16, 23, 145, 153 |
| Belatrix | 87 |
| Betelgeuse | 87 |
| Big Dipper | 50 |
| Boötes | 100 |
| Cancer | 106 |
| Canis Major | 89 |
| Canis Minor | 89, 90, 91, 92 |
| Capella | 86 |
| Capricorn | 79 |
| Cassiopeia | 57 |
| Celestial Equator | 13 |
| Celestial Horizon | 16 |
| Celestial Navigation | 5 |
| Celestial Poles | 12 |
| Celestial sphere | 12 |
| Centaurus | 108 |

| | |
|---|---|
| Cicumpolar | 49 |
| Circumferernce of Earth | 118 |
| Civil Day | 129 |
| Civil Twilight | 136 |
| Constellations | 47 |
| Co-ordinated Universal Time | 134 |
| Corona Borealis | 101 |
| Crux | 110 |
| Cygnus | 70 |
| Dark Side of the Moon | 24 |
| Declination | 13, 146 |
| Deneb | 70 |
| Denobola | 54 |
| Dubhe | 51 |
| Dwarf Planets | 35 |
| Earth Dimension Table | 119 |
| Ecliptic | 12 |
| Ecliptic Poles | 12 |
| El Nath | 81, 86 |
| Enif | 66 |
| Equation of Time | 125 |
| Equator | 13, 116 |
| Equatorial Circumference | 118 |
| Equinoxes | 13 |
| Evening Star | 36 |
| First Point Of Aries | 19, 71, 76, 93, 127 |
| Formulhaut | 77 |
| Gacrux | 111 |
| Gas Giants | 35 |
| GD | 135 |
| Gemini | 56 |
| Geographic Mile | 117 |
| Geographic Poles | 12 |
| GMT | 132 |
| Great Bear | 50 |
| Great Circle | 114 |
| Greenwich Date | 135 |
| Greenwich Hour Angle | 18 |
| Greenwich Mean Time | 132 |
| Hadir | 108 |
| Hamal | 94 |
| Hoizontal Parallax | 32 |
| Inner Planets | 35 |

| | |
|---|---|
| Jupiter | 42 |
| Kaus Australis | 95 |
| Kilometre | 117 |
| Kochab | 52 |
| Latitude | 114, 116 |
| Leo | 54 |
| LHA | 16, 130, 141 |
| Libra | 104 |
| Little Dipper | 51 |
| LMT | 130 |
| Local Hour Angle | 16, 130, 141 |
| Local Mean Time | 130 |
| Longitude | 115 |
| Lower Limb | 28 |
| Lyra | 68 |
| Magnetic Poles | 12 |
| Markab | 66 |
| Mars | 39 |
| Mean Circumference | 118 |
| Mean Noon | 129 |
| Mean Solar Day | 129 |
| Mean Solar Time | 125 |
| Mean Sun | 125 |
| Measurement Conversion Table | 117 |
| Menkent | 108 |
| Meridians | 115 |
| Meridional Circumference | 118 |
| Mirfak | 62 |
| Moon | 24 |
| Morning Star | 36 |
| Nadir | 15 |
| Nautical Mile | 117 |
| Nautical Twilight | 49, 136 |
| Navigational Planets | 35, 140 |
| Navigational Stars | 47, 140 |
| Neap Tides | 27 |
| North Star | 52 |
| Northern Crown | 101 |
| Nunki | 95 |
| Oblate Spheroid | 118 |
| Observed Altitude | 16, 31 |
| Ocean Tides | 26 |
| Orbit | 12 |

| | |
|---|---|
| Orion | 87 |
| Outer Planets | 35 |
| Parallax | 31 |
| Pegasus | 65 |
| Perseus | 60 |
| Phases of the Moon | 25 |
| Pisces | 71 |
| Piscis Austrinus | 76 |
| Planets | 34 |
| Pleiades | 83 |
| Plough | 50 |
| Pluto | 34 |
| Pointers | 52, 111 |
| Polar Circumference | 118 |
| Polaris | 52 |
| Pole Star | 52 |
| Pollux | 57 |
| Procyon | 91, 92 |
| Prograde Motion | 41 |
| RA | 19 |
| Rational Horizon | 16 |
| Reference Lines | 48 |
| Refraction | 31 |
| Regulus | 54 |
| Retrograde Loop | 43 |
| Retrograde Motion | 41 |
| Rigel | 87 |
| Rigel Kentaurus | 108 |
| Right Ascension | 19 |
| Rotation | 12 |
| Rotational Velocity | 13 |
| Sagittarius | 95 |
| Sailing Stars | 83 |
| Saturn | 44 |
| Schedir | 59 |
| Scorpius | 97 |
| Semi Diameter | 30 |
| Sensible Horizon | 16 |
| SHA | 20 |
| Shaula | 97 |
| Sidereal Hour Angle | 20 |
| Sirius | 89 |
| Sky Maps | 43 |

| | |
|---|---|
| Small Circle | 114 |
| Solstices | 14 |
| Southern Cross | 110 |
| Southern Triangle | 92 |
| Spherical Trigonometry | 158, 164 |
| Spica | 102 |
| Spring Tides | 27 |
| Square of Pegasus | 66 |
| Standard Time | 134 |
| Star Maps | 48 |
| Stars | 47 |
| Statute Mile | 117 |
| Summer Solstice | 14 |
| Summer Triangle | 67 |
| Taurus | 81 |
| Terrestrial Planets | 35 |
| Time | 124 |
| Tropic of Cancer | 14, 107 |
| Tropic of Capricorn | 14, 80, 97 |
| True Altitude | 16, 32 |
| True Sun | 124 |
| Twilight | 135 |
| Universal Time | 133 |
| Upper Limb | 28 |
| Ursa Major | 50 |
| Ursa Minor | 51 |
| UT | 133 |
| UTC | 134 |
| Vega | 69 |
| Venus | 36 |
| Vernal Equinox | 13, 76 |
| Virgo | 102 |
| Visible Horizon | 16 |
| Where To Look Method | 140 |
| Winter Solstice | 14 |
| Winter Triangle | 92 |
| Zenith | 15 |
| Zenith Distance | 15, 22 |
| Zodiac | 48 |
| Zone Time | 134 |
| ZT | 134 |

Printed in Great Britain
by Amazon.co.uk, Ltd.,
Marston Gate.